The Types, Properties, and Applications of Conductive Textiles

The Types, Properties, and Applications of Conductive Textiles

By

Sandra Varnaitė-Žuravliova

Cambridge
Scholars
Publishing

The Types, Properties, and Applications of Conductive Textiles

By Sandra Varnaitė-Žuravliova

This book first published 2020

Cambridge Scholars Publishing

Lady Stephenson Library, Newcastle upon Tyne, NE6 2PA, UK

British Library Cataloguing in Publication Data
A catalogue record for this book is available from the British Library

ISBN (10): 1-5275-4175-4
ISBN (13): 978-1-5275-4175-7

CONTENTS

ABBREVIATIONS

ESD – electrostatic discharge

PPE – Personal protective equipment

EMSE – electromagnetic shielding effectiveness

EMR – electromagnetic radiation

RFR – radio frequency radiation

EMC – electromagnetic compatibility

RF – radio frequency

AC – alternating current

DC – direct current

SE – shielding efficiency

EMS – electromagnetic shielding

ICP – inherently conductive polymers

EMI – electromagnetic interference

ITS – intelligent textile system

MPD – modal power distribution

INTRODUCTION

The vertical resistance of humans as conductors is quite low, so electrostatic charges can accumulate on the garments of humans, who are insulated from the earth. Many electronic components can be damaged by electrostatic discharge or accumulated electrostatic charges can be transferred from the body to an electronic device by touching, thus damaging it. Electrostatic charges can be accumulated on the surface of the fabric due to tribo-phenomenon or due to induction charging. In both cases, the process of charge accumulation is not stable. The process depends on the intensity of the appropriate effect (rubbing or induction charging treatment), atmospheric (especially moisture) conditions and time factor.

Previous studies have shown that the electrostatic properties of each fibre are different and resistance values also depend on atmospheric conditions. The important characteristics of conductive fabrics are fibre content and the structure of the fabric.

Therefore, it is particularly necessary to control undesired static electric charge in those places, where a flammable or explosive medium might exist. In such cases, humans have to be grounded directly or through conductive footwear. Protective clothing is designed so as to prevent or reduce skin burn from incendiary discharge. According to the field of application, protective clothing must also fulfil specific requirements.

The application fields and forms of conductive textiles are very wide. Conductive materials can help to avoid charge accumulation on a device or humans, and also protect from incendiary discharge or electromagnetic waves at frequencies that are potential hazards to health. Conductive textiles are also utilized as sheet covers for equipment or to shield a space from electromagnetic fields. They are also used to ensure the closed current circuit needed for Smart or e-textiles.

Although conductive textiles are typically produced not only as shields against charge dissipation and EMI, they are also used in other specialist applications such as sensors, antennas, flexible heaters and specialized apparel. Electrically conductive woven or knitted fabrics with particular electrical properties offer an opportunity to achieve required EMI shielding effectiveness in various frequency ranges. Moreover, these thin shielding materials can provide the additional benefits of being user-friendly, and able to be used on surfaces of all shapes because of their structural order and ability to flex.

Various techniques are used to improve the conductivity of textiles: introduction of electrically conductive yarns (carbon fibres, metal fibre); metallization of fabrics or yarns (voltaic, vacuum vaporization); lamination or coating of conductive layers onto the fabric surface with metal particles, transparent organic metal oxides, carbon or intrinsically conductive polymers (ICPs).

The different chapters in this book provide basic knowledge about the principles, roles, types and evaluation methods of anti-static and conductive textile materials, which are used for protection against charge dissipation, incendiary discharge, intense electrostatic field and electromagnetic interference (EMI) at specific frequencies. The basic properties of different types of conductive fibres/filaments and the manufacturing processes of conductive textile products will also be discussed.

1. ELECTROSTATIC PHENOMENON

Humans face static electricity every day in various surroundings. Most people are mainly affected by an extremely low-frequency (50–60 Hz) electromagnetic field. Many electrical effects are harmless and imperceptible, but static electricity can also lead to very dangerous situations: ignition and explosion, electric shock, when there are other hazardous substances in the environment that can cause injury or death to a person (Nurmi et al., 2007; Kowalski and Wróblewska, 2006; Lerner, 1985; Lambrozo, 2001).

Static electricity is treated as a set of phenomena associated with the formation and accumulation of electrostatic charges on materials with low electrical conductivity and on conductive objects isolated from the ground. Electrostatic charges are created as excess electrical charges.

Many electronic components are damaged from electrostatic discharge (ESD). This discharge can be avoided if components are handled in an ESD-protected area (EPA). Electrostatic fields and sources of ESD are controlled in the EPA in order to keep ESD risks to an insignificant level. Static electricity also causes operational problems during the production and transportation of materials, e.g. by causing fabric pieces to adhere to each other or by attracting dust (Nurmi et al., 2007).

An operating device can be damaged by the transfer of an accumulated electrostatic charge from the human body to the electronic device; the device does not have to actually touched. The presence of a charged human body near a functioning device may be sufficient to create a voltage potential that can damage the device. It has been found that a person does not feel ESD when the voltage potential is up to 3000–4000 V and over 5000 V ESD can cause malfunctions in semi-conductor devices (Lerner, 1985; Sweet et al., 1986).

There are three main sources of electric charge that can lead to damaging effects of the ESD (Lerner, 1985):

1. An electrified person touches the device and transfers the accumulated charge to/through the device to the ground.

2. The device is a solid capacitor plane that can accumulate an electric charge, i.e. to charge triboelectrically. The ESD pulse can cause a malfunction when in contact with the ground.
3. The device is located in an electrostatic field that is generated by an electrostatically charged object that can cause potential charge through the device and damage the device.

All electrostatic effects are caused by forces between the electric charges (Nurmi et al., 2007).

2. ELECTROSTATIC CHARGE GENERATION AND CHARGING OF TEXTILES

The effects of static charges are familiar to most people because we can feel, hear and even see the sparks as excess charges are neutralized when brought close to a grounded conductor, or a region with excess charges of the opposite polarity. The familiar phenomenon of a static "shock" is caused by the neutralization of charges. However, in the real world, there are actually many mechanisms that can lead to the formation of static charges, and some of the more common ones are listed below. Any one of these mechanisms can lead to static charging on textile materials, but in many cases, more than one mechanism may work together to generate static charges (Zhang, 2011).

2.1. Contact charging mechanism and the Triboelectric Series

The main source of electrostatic charge is the electrification of particles during contact, i.e. the electric charge is generated by rubbing two closed fabrics against each other and then separating them (triboelectric charging) (Nurmi et al., 2007; Lerner, 1985). Electrons can be exchanged between textile materials on contact: materials with sparsely filled outer orbital shells tend to gain excess electrons, while materials with weakly bound electrons tend to lose them. This can cause one material to be negatively charged and the other—positively charged (Zhang, 2011). If in the system of contacting materials, one of them is a grounded conductor, then the charge remains on the non-conductive material only. Due to the limited mobility of this type of charge, it is referred to as an "electrostatic" charge (Nurmi et al., 2007; Kowalski and Wróblewska, 2006; Lerner, 1985).

In addition to electrons, other elements or ions can also be exchanged between materials, e.g. in textiles made of acidic or basic polymers, or polymers with space charge layers, ions can be exchanged at the interface of two textile materials that are in contact. In this case, charges redistribute according to Boltzmann statics, i.e. charges move between the two contact materials in numbers (n) that depend on the activation energy ΔG (Zhang, 2011):

$$n = n_0 \exp\left(\frac{-\Delta G}{kT}\right) \tag{1}$$

where: n_0– the pre-exponential constant, k– the Boltzmann's constant, T– the absolute temperature.

The size and the sign of the resulting electrostatic charge on textiles depends on such factors as the chemical composition, the physical state and structure of the material, the type and amount of admixtures of foreign substances in the electrifying bodies and the electrical conductivity of the material (Nurmi et al., 2007; Kowalski and Wróblewska, 2006; Lerner, 1985).

In the contact-induced charge separation mechanism, the polarity of static charges generated on the materials depends on their relative positions in the triboelectric series. The triboelectric series is an empirically compiled list where materials are arranged from top to bottom depending on their relative ability to lose or gain electrons, beginning with the most positively charged material and ending with the material carrying the most negative charge (see Table 2.1.1) (Welsher et al., 1990).

According to the triboelectric series shown in Table 2.1.1 the polarity of static charge generated on a material can be predicted, e.g. when wool fibres contact with cotton fibres, wool acquires a positive charge and cotton acquires a negative charge because cotton has a better ability to gain electrons than wool. In contrast, the same cotton fibres acquire a positive charge when in contact with polyethylene fibres because cotton has a greater tendency to lose electrons compared to polyethylene (Zhang, 2011).

In many cases, rubbing two textile materials produces temperature gradients, and charges can move from a hot spot to a cold surrounding area due to the thermoelectric effect. Heating can also generate a separation of charge in the atoms or molecules of certain materials. This is so-called heat-induced charge separation (i.e. pyroelectric effect). The atomic or molecular properties of warmth and pressure response are closely connected. All pyroelectric materials are also piezoelectrics (Zhang, 2011).

Table 2.1.1. Triboelectric series of textile fibres

Material	Polarity
Asbestos	
Acetate	
Glass	+
Human hair	
Nylon	
Wool	
Fur	
Lead	
Silk	
Aluminum	
Paper	
Polyurethane	
Cotton	
Wood	
Steel	
Sealing wax	
Hard rubber	
Acetate fiber	
Mylar† film	
Epoxy glass	
Nickel, copper, silver	
UV resist	
Brass, stainless steel	
Synthetic rubber	
Acrylic	
Polystyrene foam	
Polyurethane foam	
Saran† film	
Polyester	
Polyethylene	
Polypropylene	
PVC (vinyl)	-
Teflon§ coating	
Silicone rubber	

Other static charge generation mechanisms, such as pressure-induced charge separation (i.e. piezoelectric effect), also play important roles. It is determined by the ability of textile materials to generate static charges in response to applied mechanical stress or strain. The nature of the pressure-induced charge separation is closely related to the formation of electric dipole moments in materials. Such separation is often observed in natural fibres, such as wool and silk (Zhang, 2011).

2.2. Induction charging mechanism

Charge-induced charge separation (i.e. electrostatic induction) is another mechanism for static charge generation (Zhang, 2011). During induction charging, an uncharged isolator is placed in the electrical field of the object (positively charged isolator), where the voltage of the uncharged insulator changes due to the nearby electrically charged insulator and its generated electric field (Nurmi et al., 2007). The induced charge is opposite in sign from the inducing charge (Holme et al., 1998). In non-conductive textile materials, although the electrons are bound to atoms and are not free to flow between atoms, they can move within the atoms (Zhang, 2011). Charge induction is not usually considered to be of major importance as a means of charging textiles, but it is important as a means of transferring charge from textiles to relatively conducting surfaces, such as the human body or charge-dissipating fibres (Holme et al., 1998; Jonassen, 2013).

The integrated study of Stankute et al. on electrostatic charge accumulation and kinetics on the surface of five fibre-forming polymers influenced by friction and induction charging has shown that the most significant parameters of tribocharge were determined during the contact of investigated objects with plexiglass pads. However, the values of the dynamic friction coefficient obtained using these pads were the lowest. The authors concluded that according to the results of electrostatic charge alteration (charge decay time) applying the induction charge method, all objects of investigation might be grouped into several groups: polylactide, soybean protein, cotton–Tencel–bamboo (Stankute et al., 2010).

2.3. Charging by ion or electro-bombardment mechanism

This method of charging is usually affected by creating a corona discharge, which results from raising fine points or fine wires of a conductive material, usually metallic, to a high enough electric potential to cause an electric breakdown of the local atmosphere. The lower the radius of curvature of the point or wire, the lower the potential needed for electrical breakdown to occur. If the discharging electrode is positively charged, positive ions are repelled and negative ions and electrons are attached, and vice versa. Consequently, a textile brought into the vicinity of the discharging electrode accumulates charge of the same sign as the electrode (Holme et al., 1998; Jonassen, 2013).

2.4. The surface electrode charging mechanism

Electrostatic charge can be produced on a sample by direct contact with a highly charged conductive electrode. This principle is used both as a means of charging a sample in a region directly in contact with the electrode whose charge acceptance and decay rate will be measured, and also as a means of charging a sample in a neighbouring region, not directly in contact with the electrode, to which the charge diffusion rate and/or from which the charge decay rate will be measured (Holme et al., 1998).

2.5. Electrokinetic or Zeta Potential mechanism

At a solid-liquid interface, an electrical double layer consisting of positive and negative charges is formed; one set of charges is associated with the solid phase and is fixed, the other is associated with the liquid and is mobile. As a result, there is a difference in potential between the locus of separation between the fixed charge and the mobile charge on the one hand and the bulk of the liquid on the other hand, which is known as the electrokinetic or zeta potential. Electrophoresis is the displacement of the oppositely charged layers relative to one another under the influence of an applied electrical field; the electrokinetic potential can be calculated from the electrophoretic mobility of a particle in a field of known strength. For fibres, the most important phenomenon is the streaming potential, which is the potential difference produced when a liquid is forced through a porous membrane such as a plug of fibres, or through a capillary tube. The streaming potential has been most widely used for calculating zeta potential values of fibres for use as an analytical tool for comparing the charging properties of fibre surfaces, including modified fibre surfaces, and for fundamental investigation of the effect of charging properties on dyeing processes. However, the establishment of a streaming potential between fibres and a liquid, the separation of the two phases, is a means of producing an electrostatic charge on the fibres, and extrusion of fibres through a capillary during manufacture can lead to freezing-in of the mobile charge during solidification (Holme et al., 1998).

2.6. Charge Decay mechanism

In taking steps to minimise the accumulation of charge on textiles it is sometimes possible to prevent the charging process, but often it is necessary either to shield the charge formed from locations where its influence would be deleterious, or to arrange local neutralisation of the charge within the structure, or to accelerate the processes of charge dissipation, thus reducing the maximum charge attained (Holme et al., 1998).

3. Charge Dissipation or Electrostatic Discharge of Textiles

Air is a good insulator under normal environmental conditions. However, if the electrostatic field strength reaches about 3 MV/m, the insulating property of air weakens and electrostatic discharge occurs. The type of discharge depends on different factors, among others the nature and geometry of the material in which it develops (Nurmi et al., 2007).

The corona discharge usually happens on conductors with pointed edges. The electric field increases above the breakdown field locally at the sharp surface and charge will discharge. The strength at the edge is typically about 3 MV/m (Nurmi et al., 2007).

If single polarity charges are not accumulated on a single layer of a non-conductive surface, but charges of different polarity are accumulated on two surfaces of non-conductive fabric layers of opposite charge, it is likely that a glow discharge will occur. The density of energy during glow discharge is higher than during corona discharge and may be sufficient to ignite flammable gases, liquids or powders (Nurmi et al., 2007).

Spark discharge is the best-known type of electric discharge. It happens between two conductors which have a high voltage difference between them. The best-known electrical spark discharge is lightning (Nurmi et al., 2007).

If the charges are not arranged in the form of one single layer of one polarity on a non-conducting surface but in the form of a double layer of charges of opposite polarity on the opposite surfaces of a non-conducting material in the form of a sheet, propagating brush discharges may occur. The energy density in a brush discharge is higher than in corona discharge and it may be enough to ignite flammable gases, liquids or powders (Nurmi et al., 2007).

Insulating fabrics with extremely conductive fibres are able to dissipate static charges even without grounding. Research has shown that fibres cause corona charge dissipation when highly charged. The ions thus created neutralize fabrics until their electrical potential becomes less than the corona electric charge potential at the beginning, which is a function of the conductive fibre diameter. Such fabrics (with metal fibres) are not capable of causing the spreading of combustible charge. Some fibres are designed with a trilobal core, the structure of which is designed so that corona charge dissipation occurs faster with high surface potential when high surface resistance is maintained at low voltages. This makes the fabrics safe to use in situations where there are high voltage and low surface resistances and where such environmental conditions pose a risk. However, it seems that textiles that incorporate a highly conductive mesh, formed by inserted conductive yarns, are capable of carrying the entire network charge, similar to spark charge dissipation—and therefore, for such materials, the entire surface charge—just like the charge density—is critical (Kathirgamanathan et al., 2000; Kessler and Fisher, 1997; Nelson et al., 1993; Kalliohaka et al., 2005).

If the fabric is conductive, it will retain the electrical charge until the fabric is grounded. This charge is called a mobile charge. When the electric field is created in the fabric, mobile charge carriers, i. e. positive charges will move in the field direction and negative charges in the opposite direction. If the fabric is an insulator, the electrical charge is stationary and remains until it will be somehow neutralized (Nurmi et al., 2007; Lerner, 1985).

Two types of electrostatic forces can act in particle motion: Coulomb and reflection forces. The Coulomb effect occurs when an electrified particle is carried by an electric field, and the reflection force is a polarization phenomenon that occurs when the electrically-driven particle is carried to the conductive surface. The Coulomb effect can be attractive or repulsive, and the reflection effect is always attractive (Lai, 2006).

The electric charges act on each other with forces that create an electric field. If the force acting on the charge q is F, the field strength E is defined by (Nurmi et al., 2007):

$$F = qE \tag{2}$$

In the electric fields, charges of the same polarity repel each other and charges of opposite polarity attract each other. If the charge q is a point charge at distance r, it will create an electric field E (V/m) which can be calculated using the formula (Nurmi et al., 2007; Jonassen, 2013):

$$E = \frac{q}{4\pi\varepsilon r^2} \quad (3)$$

From a small charged object the electric field strength often decreases with distance *r* (Nurmi et al., 2007).

For most materials, including the majority of textile fibres, Ohm's law of current density is valid, which describes the linear relationship between the current density *j* flowing through the fabric and the electric field strength *E* (Jonassen, 2013; Matukonis et al., 1976):

$$j = \gamma E \quad (4)$$

The coefficient γ, which is different for various materials, is called the specific electrical conductivity. The parameter ρ, which is inversely related to the specific electrical conductivity, is called the specific resistivity of the fabric (Jonassen, 2013, Matukonis et al., 1976) and vice versa—the parameter, inverse to the resistivity of the fabric, is called the specific electrical conductivity of the material:

$$\rho = \frac{1}{\gamma} \quad (5)$$

In total, Ohm's law states that the current *I* (A) between two points of the conductor is proportional to the voltage or potential difference *U* (V), between the two points and reciprocally proportional to the resistance of the conductor:

$$I = \frac{U}{R} \quad (6)$$

For a metallic conductor, *R* is independent of the applied voltage, which means that the electrical characteristic (*I* as a function of *U*) is linear. This characteristic makes it possible to distinguish conductors from semiconductors.

If the electrical resistance of the fabric is low enough, any charge accumulated on the surface of the fabric can be dissipated through the fabric surface or perpendicularly through the fabric, thereby reducing the density of the existing charge, or grounding through a particular connection. Some materials used to control static charge (electrification) have inherent conductive properties. There are three phenomena that can cause static discharge and the charge on heterogeneous dissipating fabrics is neutralized (Nurmi et al., 2007; Tappura and Nurmi, 2003; Gasana et al., 2006; Jonassen, 2013):

1) If the material is grounded, the conductive charge on or near the conductive element will be brought to the ground.
2) The charge on the insulating substrate induces a charge of opposite polarity on the grounded conductive yarn leading to partial neutralization of the whole charge. This phenomenon can also be understood as an increase in the vertical resistance caused by the grounded yarns which charge potential is lower. The charge of one polarity (in contrast to the base fabric) stays on the conductive yarn, while the charge of the opposite polarity is brought to the ground. In other words, by bringing a grounded conductor closer to the charge on the substrate fabric, its conductivity increases, and hence its potential decreases. The inner structure of the fabric, especially the distances between the grounded conductive yarns, has the greatest influence on the potential of the fabric during induction.
3) Partial neutralization of charges on the substrate fabric may also occur due to air ions formed during corona discharge when the large corona field strength is created in one place. Density and shape of the thread diameter will have a major impact on the high corona field strength. The corona mechanism does not require grounded yarns, but the surface charge density must be large enough to initiate the corona discharge.

The only phenomenon out of three, associated with the specific resistance of the fabric is conductivity. Conductivity and induction depend on the grounding of conductive yarns.

The electrostatic properties of the materials determine the following parameters to the greatest extent (Jachowicz, 2013):

1) Leakage resistance R_u, i. e. the resistance of the path over which leakage current flows, which primarily determines the possibility for the accumulation of an electrostatic charge on the material. This refers to the total electrical resistance, measured between the surface of the object and the ground. It is, therefore, a transition resistance to the ground, the value of which, in addition to the conductivity of the material, is also affected by the resistance of separating it from ground construction materials. An electrostatic

charge cannot be accumulated on objects where leakage resistance fulfils the condition $R_u<10^6\ \Omega$.

2) Permittivity – the ability of a material to produce and maintain an electrostatic charge. Knowledge of the relative permittivity ε facilitates an approximate assessment of the expected electrification of a given material. In particular, the degree of its static electricity charge, achieved in contact with different materials, is greater the bigger the difference between the electrical permeability of this material and the permeability of the material in contact with it.
3) The relaxation time of an electrostatic charge τ determines the rate of removal of the electrified material or object. This is the time during which the degree of static electricity in the material is reduced to about 27 % of the initial value of the generated charge. It can be expressed as the product of the permittivity $\varepsilon_0\varepsilon$ and vertical resistance R_v of a material ($\tau=\varepsilon_0\varepsilon\varsigma_v$) or the product of leakage resistance R_u and electric capacity C if the electrostatic charge is accumulated in the isolated form with a ground conductive object ($\tau=R_uC$). The above takes place when a loss of charge takes place through leakage resistance R_u, a situation which is not taken into account when the process of discharging conditions plays a dominant role e.g. by the depolarization of a material or the desorption of ions. It is accepted, that the total disappearance of electrostatic charge takes place after the passage of the so-called time of complete discharge t_w ($t_w=5\tau$).

The electrical conductivity of materials, which plays a decisive part in maintaining the electrified state of an object, is expressed by the value of vertical resistance R_v and surface resistance R_s.

The surface resistivity (ρ) of a fabric is the surface resistance (R_s) (Ω) between the opposite edges along the fabric surface (EN 1149-1: 2006). The vertical resistance (R_v) (Ω) is the electrical resistance perpendicular to the surface of the material (EN 1149-2: 2000).

The electrical resistance of a piece of material is commonly measured with an Ohmmeter.

The electrical resistance R (Ω) of a conductor of length L (m) with cross section A (m^2) made out of a conductive material with electrical resistivity ρ is expressed by Pouillet's Law.

$$\rho = \frac{\mathrm{RA}}{\mathrm{L}} \tag{7}$$

The cross section of yarn is considered as the sum of the cross sections of the electroconductive filaments or fibres. Therefore, the conductivity of conductive yarns is expressed as a resistance for a given length expressed as a linear resistance (Ω/m). This value is obtained by measuring the resistance of a certain length of yarn, without taking the cross section into consideration.

In general, materials that become electrified share the following characteristics (Jachowicz, 2013):

- small electrical conductivity, for which vertical resistance is $R_v>10^4$ Ωm or surface resistance is $R_s>10^7$ Ω,
- conductivity, for which vertical resistance $R_v\leq10^4$ Ωm or surface resistance is $R_s\leq10^7$ Ω, and materials are isolated from the ground with a non-conductive material, for which vertical resistance is $R_v>10^7$ Ωm or surface resistance is $R_s>10^{10}$ Ω.

For permanent electrification to occur, vertical resistance of $R_v>10^7$ Ωm or surface resistance $R_s>10^{10}$ Ω must be present; the electrification of such materials generally results in disturbances in the surrounding environment or production processes have been carried out with their participation (Jachowicz, 2013).

Materials of vertical resistance 10^4 Ωm$<R_v\leq10^7$–10^8 Ωm or surface resistance 10^7 Ω$<R_s\leq10^{10}$ Ω generally show a slight capability of electrification and in contact with the grounded, conductive elements of the production equipment, quickly lose their generated charge (Jachowicz, 2013).

Materials with vertical resistance $R_v\leq10^4$ Ωm and surface resistance $R_s\leq10^7$ Ω are considered to be conductive, i. e. unable to accumulate an electrostatic charge, under the condition that they are not isolated from the ground with non-conductive materials (Jachowicz, 2013).

Electrostatic discharge is dangerous, when its energy W_w reaches the value of the so-called minimum energy of ignition $W_{z\min}$ of combustible material. It is possible to be within the range of this discharge, i. e. when $W_w\geq W_{z\min}$, where W_{zmin} is understood as the lowest energy of electrostatic discharge, which in determined conditions is still sufficient to cause ignition of a given combustible or explosive medium (Jachowicz, 2013; Jonassen, 2013).

Energy W, which occurs during the charge dissipation from the non-conductive fabric surface, depends on the total amount of charge carried by the electric current during the charge dissipation and from the electrostatic material, and the arrangement of the dissipation system (adjacent grounded objects). The energy generated by the dissipation of the charge can be calculated by the formula (Kacprzyk and Mista, 2006):

$$W = \frac{(V_A - V_B)Q}{2}; \qquad (8)$$

where: Q– charge, carried during the dissipation; V_A and V_B– mean surface potential before and after charge dissipation, respectively.

Where the fabric is laid on a conductive and grounded surface, formula (8) can be rewritten as follows (Kacprzyk and Mista, 2006):

$$W = \frac{\varepsilon_0 \varepsilon S}{2d} V_A^2; \qquad (9)$$

where: ε_0– electrical conductivity of the free environment (8.855×10^8 F/m); ε– specific electrical conductivity of the fabric; d– thickness of the fabric; and S– the surface area of the fabric where the charge dissipation occurred.

This formula is derived for flat fabrics, assuming there is no air gap between the fabric and the grounded conductive surface, and that the surface potential after charge dissipation V_B is zero (Kacprzyk and Mista, 2006).

D. Montgomery and colleagues have investigated the influence of filament diameter on charge transfer among filaments during rubbing when the environment and rubbing conditions were controlled. Filaments of various diameters and mean specific resistivity (polyamide, about 10^{12} Ω/cm) were rubbed with filaments with lower specific resistivity values (tantalum, about 10^{-5} Ω/cm), and with filaments with higher specific resistivity values (polyethylene, about 10^{15} Ω/cm), under different perpendicular rubbing forces. When the polyamide was rubbed with tantalum, the transferred charge was proportional to the square root of the filament diameter and perpendicular to the rubbing force between the filaments. When the polyamide was rubbed with polyethylene, the transferred charge was proportional to the square of the perpendicular rubbing force and was almost independent of the diameter. These findings formed part of the hypothesis that the transferred charge accumulates only at the point of filaments contact (i. e. at the point of the material above which the molecules penetrate deep into the material during the rubbing), on the low conductivity object (Montgomery et al., 1961).

Hersh and Montgomery, (1956), noted that when the rubbing speed between the yarns increases, two factors were noticed: the time of charge flow through the spacing between yarns and leakage from the point of contact decreases; the temperature at the rubbing point increases significantly, but it did not affect the results of the study.

4. ELECTRICAL CONDUCTIVITY AND ELECTROSTATIC SHIELDING OF TEXTILES

The electric conductivity scale of solid materials in Figure 4.1 demonstrates that the conductivity values of the most commonly used textile fibres fall into a region below 10^{-7} S/m, which corresponds to the best insulators. The most commonly used conductive materials, such as metal, show much higher values: from 10^{7} S/m for steel to 10^{9} S/m for copper and silver (Marchini, 1991).

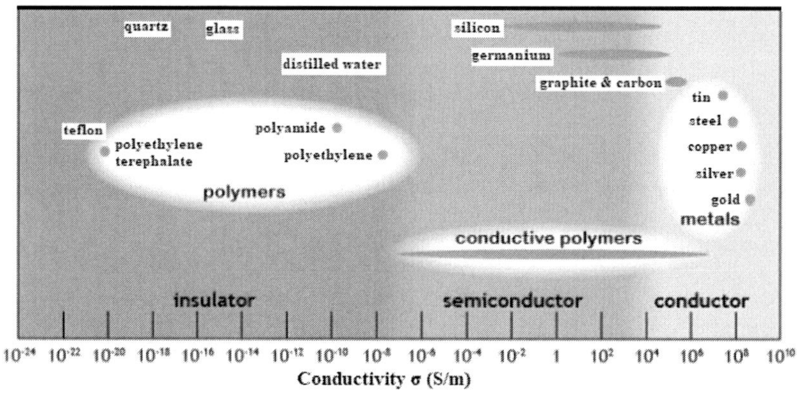

Figure 4.1. The electrical conductivity of materials

The surface resistivity of textiles can be divided into such groups (Lin and Lou, 2003; CEN/TR 16298: 2011):

1) EMI/RFI shielding materials: less than 10^{4} Ω;
2) conductive textiles: less than 10^{6} Ω;
3) static dissipative materials: from 10^{6} to 10^{12} Ω;
4) anti-static textiles: from 10^{10} to 10^{12} Ω;
3) insulation textiles above 10^{12} Ω.

Many synthetic fibres used in the production of textiles are insulators with a specific resistivity of about 10^{15} Ω. This is much higher than the materials used for electromagnetic shielding materials. For example, the best quality anti-static-electrostatic clothing must have a specific surface resistivity of between 10^5 Ω and 10^9 Ω (Lin and Lou, 2003). But, it is enough for the anti-static-electrostatic fabrics to have a specific resistivity of between 10^9 and 10^{13} Ω and for static charge-dissipating fabrics between 10^2 and 10^6 Ω, and for shielding materials less than 10^2 Ω (Chen et al., 2007).

Because simple polymeric materials are electrically non-conducting, the surface resistance of such materials is usually higher than 10^{12} Ω, so electrons may easily accumulate on the polymer surface. Such accumulated electrons create a high voltage in a short time, which can destroy mechanical elements or even cause an explosion. Processes for the production of conductive polymers can be divided into two groups: processes by which the polymer itself is produced as a conductor and where conductive particles (metal powders, fibres, etc.) are inserted into the polymer matrix in the manufacturing process (Lei et al., 2004).

For materials with some electrostatic shielding effects, the E_R measured according to EN 1149-3: 2004 Method 2 is less than E_{max}. Occasionally, a transition peak appears in curves drawn by the recorder. Such peaks are not taken into account when calculating E_R. If $E_R < E_{max}/2$, this is recorded as $t_{50} < 0.01$ s. If the field strength displayed in 30 seconds does not decrease to $E_{max}/2$, this is recorded as $t_{50} > 30$ s.

The shielding effect of the test fabrics is not sudden, so the results obtained during the test EN 1149-3: 2004 (induction charging method) can be divided into three types: metal, core and homogeneous (see Figure 4.2). Fabrics, whose curves are drawn without any initial spike, are classified as metal. The core material curves have an initial spike E_R which quickly disappears (30-50 μs) to value $E_{max}/2$. If there is no shielding factor for the fabrics, the E_R is equal to E_{max} and fabrics behave like insulators (EN 1149-3: 2004; Paasi et al., 2004).

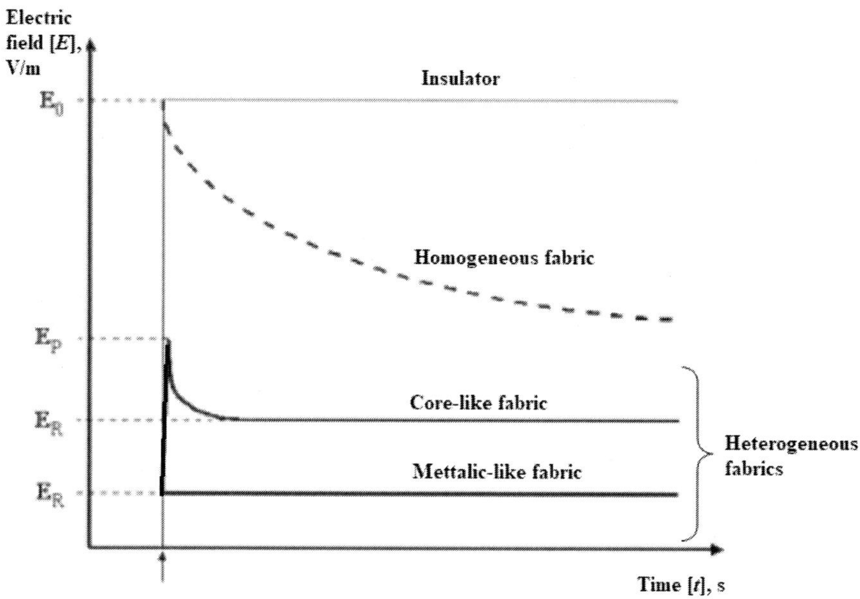

Figure 4.2. Shielding effect of different textile fabrics (Paasi et al., 2004)

The charge-discharge variation can be calculated using the following equation (Vogel et al., 2006):

$$E = E_{\max} e^{\frac{-t}{RC}} \tag{10}$$

where: E_{max}– the strength of the electrostatic field without a sample (initial field strength); E– a variation of electrostatic field strength in time; C– capacitance of the measuring device; R– resistance of the object under investigation; t– time.

Typically, there is a good relationship between homogeneous fabrics, with surface resistivities of less than 10^{11} Ω, and a duration of the charge decay (EN 1149-1: 2000).

Vogel et al. conducted measurements of homogeneous and heterogeneous fabrics with the ICM-1 charge decay device, according to EN 1149-3, method 2. They found that polyester and polyester/cotton fabrics do not have any shielding properties, but the electric charge decay time is lower for multifiber fabric than that of the man-made fabric, and with the incorporation of conductive fibres into these fabrics the charge decay properties significantly improve (Vogel et al., 2006).

The vertical resistance can be an important feature by itself or to complement the specific surface resistivity of the clothing fabric. The low vertical resistance (e.g. less than 10^8 Ω) of the electrostatic charge dissipative clothing, is a useful property to reduce the surface resistivity. However, it is often impossible to rely on this useful property, as the isolation clothing worn under the upper clothing may interfere with the upper clothing's contact with the skin while preventing the electrostatic charge decay directly through the body. For special purposes, such as arc welding protective clothing (when voltage is typically less than 100 V), higher vertical resistance (e.g. greater than 10^5 Ω) may be required to ensure proper insulation (EN 1149-2: 2000).

Polyaniline in its thick form is electrically conductive and soluble in organic solvents such as toluene and xylene. Polyaniline solutions can be used in the finishing of filtration materials in order to improve their water removal properties. Compared to conventional anti-static filtration materials, polyaniline-treated polyester materials are easily converted into various types of filtering materials and bags. Usually, the conductivity values of polyaniline-treated materials are between $10^4 \div 10^9$ S/cm (Järvinen and Puolakka, 2003; Kuhn, 1997; Rivas and Sanchez, 2001; Molina et al., 2009).

Air filters, produced from fibres, are widely used for dust collection and environmental protection. Effective filtering of submicron particles is very important as these particles pose a health threat. Fibrous materials used for air filtration provide high filtration efficiency, low air resistance (because of low-pressure differential across the filter) and show good dust collection efficiency. Fibres in the filter are constantly charged, and electrostatic charges enhance filtration efficiency compared to mechanical filters (Motyl and Lowkis, 2006).

The efficiency of collecting dust particles from electric filters depends on many parameters, such as the density of the fibre charge, the tightness of the filter, the thickness of the filter and the size and charge of the aerosol particles. The efficiency of the collecting increases linearly with the increase of the quantity of the electrical components. Solid aerosol particles block the pores of the electric filters, and the deposited aerosol reduces the electrical charge on the surface. The liquid aerosol covers the surface of the charged fibre and reduces surface charge and collecting efficiency. Increased specific humidity can reduce the efficiency of the filters. In their work (Motyl and Lowkis, 2006), researchers studied surface tension of polypropylene non-woven fabrics according to Reedyk and Perlman alignment method (Perlman and Reedyk, 1968). The researchers have found that polypropylene non-woven material with a higher surface density and thickness has better charge accumulation and water repellency properties (Motyl and Lowkis, 2006).

Michalak and students (Michalak and Bilska, 2002) have researched non-woven fabrics made of polypropylene and electrically conductive fibres. The vertical and surface resistances of non-woven fabrics and fibres were investigated; the ability of fibres and fabrics to electrify were also assessed during the work. The results showed that even a small amount (0.5 %) of electrically conductive fibre in the fabric reduces the vertical and surface resistances at almost 10^7 degrees. Mixtures of non-woven fabrics have been described as having the lower electrical capacity.

Plastics with conductive coatings or metal fibres embedded during the forming stage are increasingly used. Recently, attention has been paid to lightweight and flexible materials such as textiles coated with absorbent material. These materials are flexible and inexpensive and have good device and human protection properties against electromagnetic radiation (Nurmi et al., 2007; Pietranik and Stawski, 2004).

4.1. Influence of specific humidity on the electrostatic properties of fabrics

The dependence of specific electrical resistivity on the specific humidity of the environment is higher for fabrics with higher moisture absorption than for fabrics with lower moisture absorption properties (Tappura and Nurmi, 2003; Ramer and Richards, 1968; Rizvi et al., 1998).

Problems arising from the accumulation of electrostatic charge on fibres, fabrics and other textiles are very diverse. The biggest problems arise with fibres with quite low moisture absorption at low relative humidity, as even moisture-absorbing fibres can cause permanent static charging in such conditions. Since conventional textile fibres are insulators or have a sufficiently low electrical conductivity, it is very difficult for the accumulated charge to dissipate. Electrostatic charge is a surface phenomenon where textile fibres have an extremely high surface to volume ratio. The sensitivity of the fibres to electrostatic charge is high enough. Sensitivity is greater the lower the humidity of the environment. Fibres with sufficiently high moisture absorption, such as wool and cotton, usually do not generate a high electrostatic charge at average ambient humidity, but at low relative humidity, they tend to generate an electrostatic charge, just like fibres with low moisture absorption. Under dry environmental conditions, cotton has a relatively high surface resistance and hardly dissipates the accumulated charge. Fabric charging is not a big problem at high relative humidity. Problems arise in the textile production process and in the processes where charges can accumulate in the processing of textiles and in other parts of the textile chain of logistics, such as transportation (Nurmi et al., 2007; Gonzalez, 2005; Knittel and Schollmeyer, 2009).

Onogi et al. (1996) investigated charge decay properties of textiles such as wool, cotton, polyamide, and polyester electrolyzed by the triboelectric method. They found that static charges from textiles mostly dissipate through electrical conductivity to the ground. However, some of them can also dissipate into the air by evaporating water molecules. Their work confirmed atmospheric electrical charge dissipation into the air by water molecules during evaporation from the fabric surface resulting in reduced static charge. The velocity constant of the dissipation of the atmospheric charge depends on the amount of water in the textiles. The authors obtained a linear relationship between the velocity constant and the amount of free water above the critical water content in the fabric. The water molecules absorbed by textiles firmly attach to the polymer molecules or exist free in textiles. The so-called "free water" plays an important role in the dissipation of the atmosphere triboelectric charges. Similar results have been obtained by other authors exploring the influence of water content on the velocity of dissipation on the fabric surface (Nakamura et al., 1981; Nakamura et al., 1983; Onogi et al., 1997).

Onogi et al. (1997), clarified the conclusions described in his earlier works. He found that charge dissipation speed has a linear dependence on the amount of free water that is above the critical water content in the fabric at only 20 °C. However, they found no correlation between the velocity of charge dissipation and the amount of free water at other temperatures, i. e. molecules, which are in textiles, cannot be divided into only two types. It seems that the velocity of charge dissipation depends on the amount of water in the fabric and the specific humidity in the environment.

Lutz and Kinbersberger investigated the dependence of epoxy resin charge dissipation properties on specific humidity. They found that the specific surface resistivity significantly decreases from 10^{16} to 10^{13} Ω degrees when the relative humidity increases. The highest jump in the specific surface resistivity is visible between 60 % and 80 % relative humidity. The time decay is also shortened as the relative humidity of the environment increases (Lutz and Kindersberger, 2009).

Ramer and Richards (1968) also found that the charge decay time is slower for fabrics with higher surface resistivities. Research has shown that the lowest specific surface resistivity and the fastest charge decay was found for polypropylene fabric, compared to other investigated fabrics of different fibre content. Thus, there is a strong tendency for the velocity of charge decay to correlate with the specific resistivity

The increase in relative humidity reduces unwanted static factors for many types of fibres, although Sereda and Feldman (1964) have described that some fibres can create higher static charges at higher specific humidity than at lower specific humidity. Because the moisture of the fibres significantly affects the electrical properties, the specific humidity has a significant influence on the static electrification of the textile materials. They noted that the maximum electric charge which may accumulate when textiles are exposed to continuous friction when the monomolecular water layer theoretically exists on the fibre surface is reduced.

Gonzalez et al., (2001) reported that the ordinary standard atmosphere for testing textile materials—65 % specific humidity—is too high to carry out static electricity tests, because there is no significant difference between the values of the different fibre resistances, however, under lower relative humidity, the difference between the values obtained is more prominent. Therefore, he recommended performing tests at 20–30 % specific humidity.

Paasi et al., (2001) investigated the specific surface resistivity and charge dissipation parameters of various textile materials used for personal protection and packaging. Studies have shown that the focus should be on the management of ESD with a specific environmental humidity of less than 20–30 % since under such conditions some ESD protective materials have a very limited ability to dissipate the accumulated charge and may become insulators. Researchers have found that it is very important to select fabrics for ESD control purposes very carefully, at low specific humidity values. It should be kept in mind that choosing fabrics, with a specific surface resistivity close to 10^5 Ω, may be risky because fabrics with specific surface resistivity within the range of 10^9–10^{11} Ω provide better protection during an ESD event. In accordance with IEC (International Electrotechnical Commission) standard 61340-5-1, the specific surface resistivity of the measured fabric should be less than 1×10^{12} Ω in order to control the risk of ESD.

Kan and Yuen (2008) investigated the dependence of electrical charges' half decay times and moisture content of polyester fabrics. They found that an increase in the moisture content of polyester fabrics shortens the charge half decay time, i.e. anti-static properties of polyester fabrics significantly improve. Since moisture consists of water molecules and is polar by nature, the conductivity of polyester fabric at higher humidity quantities improves. The local static charge on the surface of the polyester fabric dissipates much easier. The moisture layer formed on the surface of the fabric can dissipate in the air and at the same time carry a sufficient amount of static charge from the surface of the fabric to the air by reducing the static charge on the fabric surface. As the moisture content is inversely proportional to the charge decay time the increase in humidity of the low-temperature plasma-treated polyester fabric would reduce the charge decay time and thus would improve the anti-static properties of the polyester fabric.

Rizvi et al., (1998; 1995) investigated the electrostatic properties of thermal protective clothing in low relative humidity environments. They found that the fibre composition of the upper layer of the protective garment had a greater impact on electrical potential and energy than the inner layer. The inner layer has an influence when the charges were generated by separating, but not rubbing with another fabric. The dissipation potentials and energies were lower, however, than the outer layer. The energy released was less when the specific humidity during the test was lower. They also found that in extremely dry conditions rubbing activity, such as slipping in the vehicle seat, can cause charge dissipation energy of up to 15 mJ and can electrify a person wearing protective clothing, up to 13 kV. Separation of clothing is better than rubbing in order to induce much lower energy. Although the nature of the inner garment in the system does not affect the dissipation of the charge from the body, it has been discovered that the top layer is the most important. Researchers have found that, at low specific humidity, upper clothing made from aramid/carbon fabric provides the least potentials and energies. The maximum values were measured for fabrics, used for upper clothing, from FR-cotton and 100 % aramid. While people in the industry believe that cotton clothing has the lowest tendency to static electricity, FR-cotton has not shown better results than 100 % aramid. Anti-static clothing (aramid/carbon and aramid/stainless steel) has been described as the safest clothing in a low humidity environment. However, they may still cause a rather high electrostatic charge due to friction. For all subjects with anti-static clothing, an aramid/carbon fabric produces less dissipation energy than anti-static clothing from aramid/stainless steel material.

Hersh and Montgomery (1952; 1955) studied electrical properties of various yarns including polyamide, viscose and wool. They found that cyclic washing and long sample conditioning have an important influence on the electrical resistance of the yarns. The results published by the authors show that as the number of washing cycle increases, electrical resistance increases; for example, the electrical resistance of unbleached wool yarn is 4.2×10^{12} Ω, after one wash cycle it increases to 14×10^{12} Ω, and after three wash cycles it increases to 16×10^{12} Ω. The downward trend in resistance is seen with increasing conditioning time: after 36 hours of conditioning, the electrical resistance of the viscose filament yarn is 0.69×10^{12} Ω, and after more than 108 hours—0.65×10^{12} Ω and this value remains even after the conditioning time is extended to 156 hours.

Lei et al. (2004) investigated the dependence of the residual voltage of polyamide non-woven fabrics from different surfaces. They found that the surface structure of polyamide non-woven fabrics has a very significant impact on residual electrostatic voltage. Laminated patterned non-woven fabrics are very effective at dissipating electrons due to their increased surface area. The residual electrostatic voltage of laminated double line pattern non-woven fabrics is low compared to the residual electrostatic voltage of non-woven fabric with stainless steel.

Sweet et al. (1986) investigated the dependence of the charge decay time on the specific moisture of urethane foam coated with amino and anionic anti-static particles. They found that used chemicals migrate to the surface of the fabrics by forming a thin, conductive aqueous electrolyte surface by which the static charge leaks down. As the specific humidity increases from 4 % to 13 %, the static dissipation time decreases linearly from 3 s to 1.5 s.

5. EVALUATION METHODS OF CONDUCTIVE TEXTILES

Many methods have been designed and developed to evaluate the electrostatic properties of textiles so as to establish a pattern for their anti-static behaviour. Four major organizations that publish electrostatic test standards are the American Society of Testing and Materials (ASTN), the Electrostatic Discharge (ESD) Association, the American Association of Textile Chemists and Colorists (AATCC) and the International Standards Organization (ISO). These organizations have developed test methods to assess resistivity, and static charge generation, accumulation and decay. There are three accepted test method categories that can be used to evaluate the anti-static properties of textile materials (Zhang, 2011; Holme et al., 1998):

- The **direct method** category typically consists of measuring electrical properties such as electrical field, the potential, charge amount, or the rate of ESD after developing charges on the material by certain treatments.
- The **indirect method** generally involves the use of other indicators such as electrical resistivity or conductivity of the textile materials. Resistivity is the inability of a material to conduct electric current. Conductivity is the opposite of electrical resistance, and it refers to the ability of the material to allow current to flow when a potential difference is applied. Since the static charges are generated on the surface of textile materials, most work focuses on the measurement of the surface resistivity and conductivity. Surface resistivity or conductivity can be measured by using either concentric ring electrodes or parallel plate electrodes.
- The **simulation method** consists of methods that study the behaviour of textile materials in simulating the end-use for which they are intended, instead of actual electrostatic properties.

EN 1149-1 is a European Standard that specifies a test method for materials intended to be employed in the production of static dissipative protective clothing (or gloves) to avoid incendiary discharge. This test method is not applicable for materials to be used in the manufacturing of protective clothing or gloves against mains voltages.

The principle of the test method is that the specimen is placed on an insulating plate and an electrode assembly is rested on the specimen. A DC potential is applied to the electrode assembly and the surface resistance (Ω) of the fabric is measured. The measuring circuit is shown in Figure 5.1.

The surface resistivity is the surface resistance (Ω) between opposite edges of a square of the material along the surface of the material. It is independent of the electrode dimensions and is calculated by multiplying the measured surface resistance by an appropriate factor *k*. The factor *k* is calculated using the following equation:

$$k = \frac{2\pi}{\log_e\left(\frac{r_2}{r_1}\right)} \qquad (11)$$

where r_1– the radius of the inner electrode, mm; r_2– the inner radius of the outer electrode, mm.

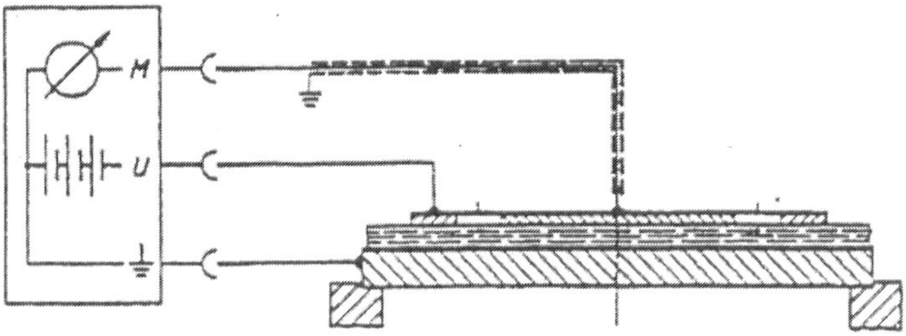

Figure 5.1. The measuring circuit of surface resistance

The specimen shall be conditioned for at least 24 h prior to testing and shall be tested in the following atmospheres: air temperature (23±1) ^{0}C, relative humidity (25±5) %. The surface resistance of materials can depend on a large degree on relative humidity. The lower the relative humidity the higher the surface resistance.

EN 1149-3 European Standard specifies methods for measuring the dissipation of electrostatic charge from the surface of materials for garments. The test methods are applicable to all materials, including homogeneous materials and inhomogeneous materials with surface conducting fibres and core fibres. The standard describes two methods for measuring the rate of dissipation of electrostatic charge of garment materials: triboelectric charging and induction charging. In both cases, the charge is monitored by observation of the electrostatic field it generates and this is done using non-conducting field measuring instruments. The principle of the first method is that test materials are charged by rubbing against cylindrical rods mounted on a vertically running slider. The electrical field strength from the charge generated on the test material is observed and recorded using an electrostatic field metre connected to a graphical recording device (see Figure 5.2). The charge dissipation of triboelectric accumulations is explained by the simple phenomenon, i.e. charges dissipate due to conductivity. The function of charge dissipation reduction is often an exponential function of time. Thus, static dissipation is characterized by the time required to dissipate the charge to half its initial value. The value of half decay of the electrostatic charge depends significantly on the condition of the fabric surface. Environmental conditions and the amount of moisture in the fabric have a significant impact on the accumulation of static charge. However, charge dissipation is not the only conduction mechanism to leak down charges accumulated on the surface of the fabric. These charges also penetrate into the fabric and also dissipate into the air (Onogi et al., 1996; Shashoua, 1958; Wilson, 1963; Guanghui et al., 1998).

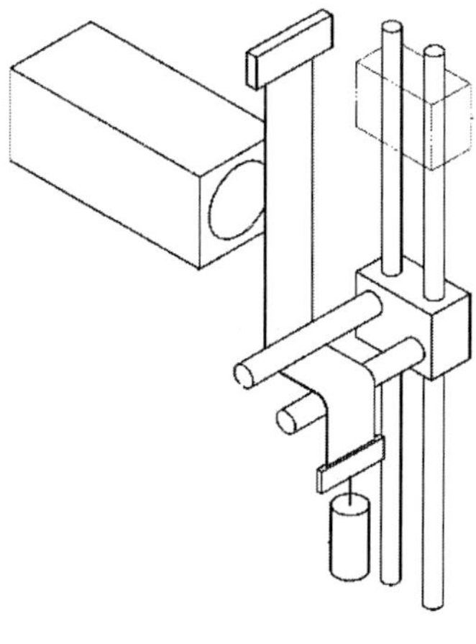

Figure 5.2. Example of equipment for triboelectric charging method

The charging of the test specimen during the second test method – induction charging – is carried out by an induction effect. Immediately under the test specimen, which is horizontally arranged, a field-electrode is positioned, without being in contact with the specimen. A high voltage is rapidly applied to the field-electrode. If the specimen is conductive or contains conducting elements, a charge of opposite polarity to the field-electrode is induced on the specimen. Field from the field-electrode which impinges on the conducting elements does not pass through the test specimen and the net field is reduced in a way that is characteristic of the material under test. During induction charging, migration of the charge can take place at any conductive element in the material (Paasi et al., 2004). This impact is measured and registered behind the specimen with an appropriate field measuring probe. The view of the measuring device is presented in Figure 5.3.

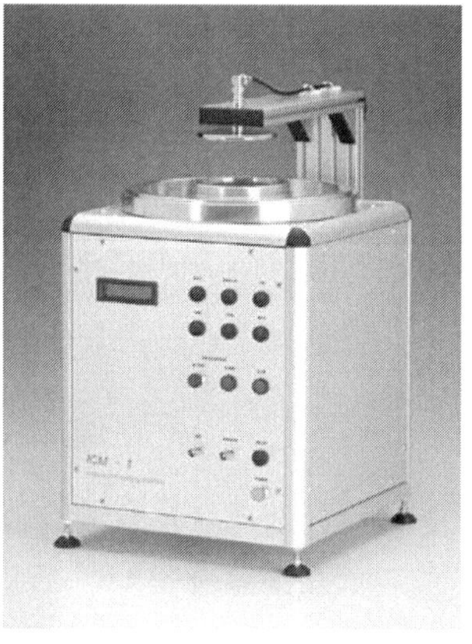

Figure 5.3. The measuring device ICM-1 (for induction charging method)

As the quantity of induced charge on the test specimen increases, the electric field registered by the measuring probe decreases. It is this decrease in the field that is used to determine the half decay time and shielding factor.

The shielding factor S is the relationship between E_{max} and E_R and calculated as (EN 1149-3: 2004):

$$S = 1 - \frac{E_R}{E_{max}} \quad (12)$$

where: E_R– maximum electric field strength indicated on the recording device with the test specimen in the measuring position; E_{max} maximum electric field strength indicated on the recording device with the test specimen in the measuring position.

In both methods, the test specimens must be conditioned for at least 24 hours in an atmosphere of (23±1) ^{0}C and (25±5) % relative humidity. Testing must be performed in the same atmosphere.

Both of these methods are suitable for evaluating the dissipation of electrostatic charges in fabrics to avoid incendiary discharges (EN 1149-3: 2004), but an electrostatic charge is generated by an electrically charged object during induction charging, rather than by direct contact between two objects, as is the case with the triboelectric method.

The induction charging method can be applied to all homogeneous and heterogeneous (made of filaments with conductive surface or core) fabrics (EN 1149-3: 2004; Paasi et al., 2004). Surface resistance measurements are not very significant for special purpose fabrics, and therefore a charge decay test has been standardized to assess such non-homogeneous fabrics for electrostatic charge dissipation garments against incendiary discharges. This method is used to accurately characterize ESD (Electrostatic Discharge) products. It provides information on the phenomenon of electrostatic shielding, but does not show what happens on the surface of the test object and inside it (EN 1149-3: 2004).

EN 1149-2 is a European Standard that specifies a test method for materials supposed to be involved in the production of static dissipative protective clothing (or gloves) to avoid incendiary discharge. This test method is not applicable for materials to be used in the manufacturing of protective clothing or gloves against network voltages.

The vertical resistance can be an important feature by itself or to complement the surface resistivity of the clothing material. The low resistance (e.g. lower than 10^8 Ω) of the clothing through which the electrostatic charge is discharged is a useful feature to reduce the surface resistivity. However, often this useful feature cannot be relied upon as the isolated clothing worn under the upper clothing may interfere with the upper clothing's contact with the skin and together prohibit the electrostatic charge from leaking directly through the body. For special purposes, such as arc welding protective clothing (when the voltage is typically less than 100 V), a high vertical resistance (e.g. greater than 10) may be required to ensure proper insulation. It should be noted that in general, the insulation properties tend to decrease with increased relative humidity.

The principle of vertical resistance tests is that electrodes are placed on opposite sides of the test sample. The difference of the DC current potential is formed between the electrodes and the resistance of the test substance is determined. The measuring circuit of vertical resistance is presented in Figure 5.4.

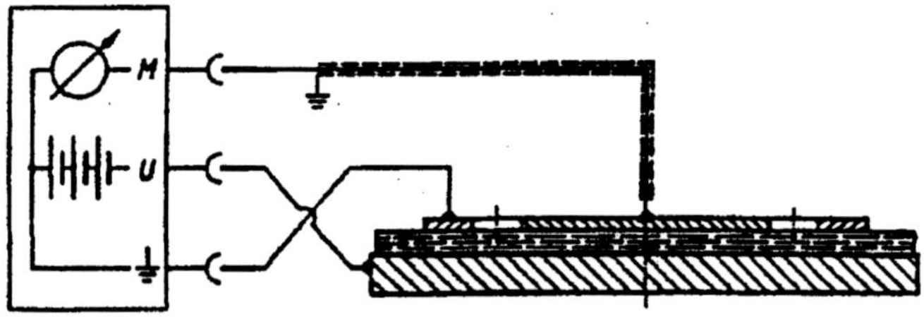

Figure 5.4. The measuring circuit of vertical resistance

The European Standard (EN ISO 11611: 2015) specifies that protective clothing worn during welding and similar processes must meet certain electrical resistance requirements, i. e. the electrical vertical resistance of all sets of clothing measured in accordance with EN 1149-2: 2000, with a relative humidity of 85±5 %, must be greater than 10^5 Ω (corresponding to a leakage current of less than 1 mA). The seam is required in prepared tested samples.

In order that the electrical characteristics of the electrically conductive yarns can be measured a suitable set-up is prepared. One end of the electrically conductive yarn is fixed on an isolated metallic clamp and the other end is curved over a metallic roller. A variable DC power source is connected to the clamp and the roller. Over these two conductive elements the yarn is set under electrical voltage (see Figure 5.5). The voltage is regulated in order to keep the electrical current lower than the critical disintegration value. If the electrical current is high, the Joule effect can lead to excessive heating of the conventional polymer yarn, which can be disintegrated through the melting mechanism of the polymer. A voltmeter is connected over a certain length L of the yarn. The electrical resistance of the yarn of length L is calculated according to the Ohm law. The yarn has a linear behaviour like every common conductor. The four point's measurement of the resistance eliminates measurement errors due to the contact resistance of the probes and the terminals of the instruments and the power source (Vassiliadis et al., 2004).

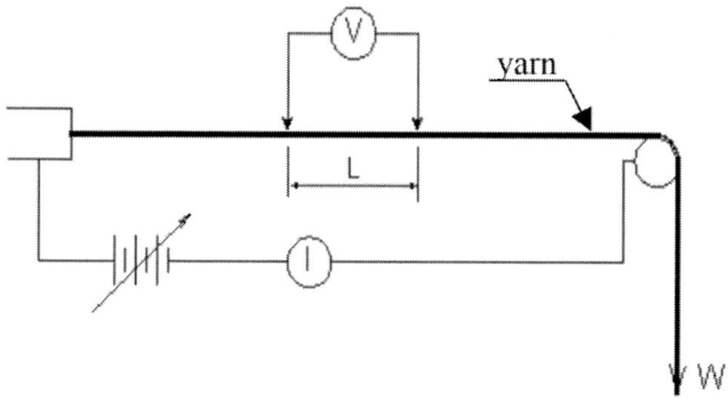

Figure 5.5. Typical set-up for DC measurements

EN 16812: 2016 is the European Standard which describes a test method for the determination of the linear electric resistance of conductive tracks for textile structures or intended for application in/to textiles, e.g. yarns, printed or coated tracks, ropes, ribbons and webbing. This European Standard is designed for materials showing ohmic behaviour and is designed for conductive tracks where electrical contact between the measurement electrodes and the conductive track is possible. This test describes a procedure to measure the linear electrical resistance of textile-based electrically conductive tracks using the measurement principle of the four-wire (four points) Kelvin method and a DC current source.

This test method can be performed as 'four-electrode–four-wire method" (see Figure 5.6) or "two-electrode–four-wire method" (see Figure 5.7).

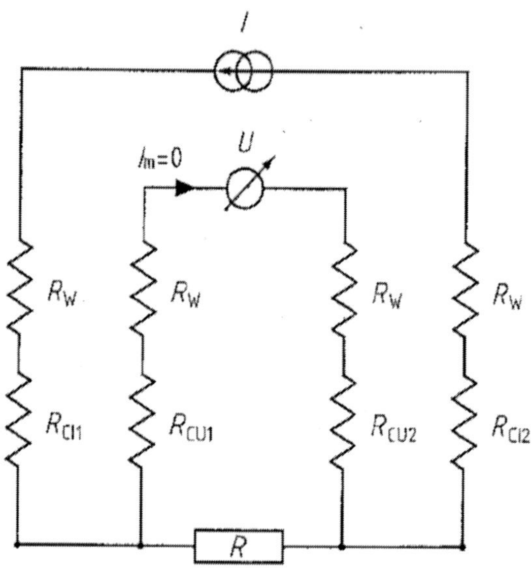

Figure 5.6. Detailed scheme for the "four-electrode– four-wire measurement"; the four electrodes (contacts) are visualized by the four nodes indicated in the scheme: I– applied current (A); U– measured voltage (V); I_m– current in the voltage measurement circuit (equivalent to zero); R_{CI1}, R_{CI2}– contact resistances in the current circuit (Ω); R_{CU1}, R_{CU2}– contact resistance in the voltage circuit (Ω); R_W– wire resistance, Ω ; R– resistance of the sample (function of electrode spacing d)=R_Lxd (Ω), identical to the measured resistance: $R=U/I$

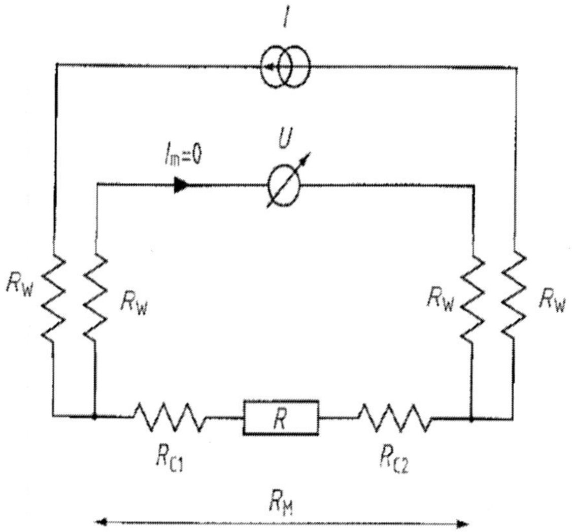

Figure 5.7. Detailed scheme for the "two-electrode–four-wire measurement"; the two electrodes (contacts) are visualized by the four nodes indicated in the scheme: *I*– applied current (A); *U*– measured voltage (V); I_m– current in the voltage measurement circuit (equivalent to zero); R_{C1}– contact resistance at electrode 1, Ω; R_{C2}– contact resistance at electrode 2, Ω; *R*– resistance of the sample (Ω); R_W– wire resistance, Ω; R_M – measured resistance, Ω, with $R_M=R_{C1}+R+R_{C2}=U/I$

By using four electrodes the contact resistance between the electrodes and the sample are compensated. R_{CI}, R_{CU} and R_W can be excluded due to the "four-electrode–four-wire measurements", so that the resistance of the specimen can be calculated by the simple formula (EN 16812: 2016):

$$R = \frac{\mathrm{U}}{\mathrm{I}} \quad (13)$$

The linear resistance R_L, Ωm, is calculated as (EN 16812: 2016):

$$R_L = \frac{\mathrm{R}}{\mathrm{d}} \quad (14)$$

where: *d*– the distance between the voltage measurement electrodes.

For "two-electrode– four-wire method" only R_W can be excluded due to the four-wire measurement so that the resistance *R* of the specimen needs to be calculated by the more complex formula (EN 16812: 2016):

$$R_M = R_{C1} + R + R_{C2} = \frac{U}{I} \tag{15}$$

The contact resistances R_{C1} and R_{C2} are usually not known. Assuming that the contact resistances are independent of the specimen length (i.e. only related to the nature of the specimen and the measurement set-up) it is possible to determine the resistance for different specimen lengths (*d*, m) and then calculate by linear resistance, R_L, Ω/m, of the specimen, using the equation (EN 16812: 2016):

$$R_M(d)=(R_{C1}+R_{C2})+R(d)-(R_{C1}+R_{C2})+R_L \cdot d \tag{16}$$

By taking into account that (EN 16812: 2016):

$$R_L = \frac{R(d)}{d} \tag{17}$$

As a result the "four-electrode–four-wire method" requires fewer measurements and calculations as compared to the "two-electrode–four-wire method". Therefore the "four-electrode–four-wire method" is the preferred method.

In cases where it is not feasible to prepare the necessary four contacts for the electrodes, "the two-electrodes–four-wire method" can be used.

6. LIMITATIONS OF RESISTANCE MEASUREMENTS FOR CONDUCTIVE TEXTILES

Due to the limitations of the surface resistivity test, heterogeneous fabrics (containing conductive filament yarns having a conductive surface or conductive core) must be measured and evaluated for the properties of incendiary discharges from different sizes of insulation surfaces, i. e. by measurement according to EN 1149-3: 2004. Electrostatic properties of homogeneous materials can also be measured according to this standard test method. The classical and most used method of measuring anti-static parameters according to EN 1149-1: 2006 is more suitable for homogeneous materials (Vogel et al., 2006; Berberi, 2001). Problems with static electricity are caused by increasing surface tension and the charge left on the fabric after it has been contacted with other fabric or was rubbed. It is possible to observe sufficiently high surface tensions on the fabric, which show low resistances and also quite low stresses on fabrics that are good insulators (Chubb, 2004).

Many chemical fibres have the increased ability to accumulate static charges and dissipate them weakly. These skills cause a lot of problems in the production of fibres and reduce the exploitation and hygiene properties of the products. It is obvious that the ability of different fibre materials to dissipate electrical charge under the same conditions is characterized by the magnitude of the specific resistivity. However, there is no reliable and sufficiently simple method to measure the electrical resistance of fibrous materials. Problems with measuring this parameter are explained by the very high electrical resistance of the polymer from which the material was made, the fibre diameter and the difficulty to ensure reliable contact between the fibres and the electrodes of the measuring device (Davydov et al., 1972).

Berberi (2001; 1998) suggested that compression properties of fibres groups have a very big influence on the electrical resistance of fabrics. The author described a new multi-step method for measuring the electrical resistance of textiles, which included the compression properties of the fibres. The specific resistivity measured and calculated by the P. G.

Berberi method describes the feature characteristics of the fibres groups and does not depend on the shape of the sample.

Standard resistance measurements are not suitable for describing the ability of a fabric to dissipate a charge on a fabric surface because (Chubb, 2004):

1. Resistance measurements indicate the fastest path of charge movement on the surface. Small values are measured when high conductivity components are considered to be a sufficiently insulating matrix. The main examples are cleaning clothes and a large number of bags containing conductive yarns with a surface conductivity. "Resistance" measurements indicate thread resistance; however, do not supply information on the ability of the charge movement among the yarns on the fabric.
2. "Resistance" measurements to divide fabrics into two groups are based on the assumption that fabrics are homogeneous and have linear current/voltage dependence. The reduction of surface resistance in many fabrics after local triboelectrification does not correspond to the exponential form that would occur with a simple "having resistance" charge motion.

Methods for measuring charge dissipation must meet the following requirements (Chubb, 2004):

- Provide results that are comparable to those of other fabrics in the practice of tribocharging.
- Measurements of the voltage generated by the electrification action are performed on the surface of the fabric and its decrease is measured over time.
- Surface voltage measurements are carried out without direct contact with the electrified area.
- Measurements must be done on the same side of the fabric that was electrified.
- Measurements are made independently of fabric construction or features.
- Minimum modifications of fabric and dissipation characteristics in the test (and this must also be applied to the triboelectric method).
- The method must be easy to construct and/or commercially available.

- It should be known how to handle the device. Instructions must be printed describing the device and giving supportive experimental measurements. All documents must be specified in the standard.

7. FABRICS AND CLOTHING CONTROLLING THE DISSIPATION OF ELECTROSTATIC CHARGES

One of the most important qualities of clothing is the comfort. It is known that the structure, composition and dimensions of the fabric, as well as the model of the clothing, have a significant influence on the comfort of wearing clothes, i.e. hygienic and thermal properties. Clothing from natural fibres provides the best comfort which artificial fibres cannot always guarantee. However, comfort properties are not very important in protective clothing. Special clothing must provide the protective properties of its purpose (Zimniewska et al., 2003). Electrostatic charge-dissipating properties are required for much protective clothing (Vogel et al., 2006).

Personal protective equipment (PPE) protecting the whole body or part of it from the impact of electric current must sufficiently insulate against the accumulation of electrostatic charge in order to protect the user in the most unfavourable conditions. Therefore, PPE components and fabrics must be selected, designed and incorporated in such a way that the electric current flowing through the protective layer is kept to a minimum and never exceeds the maximum level. PPE used in working with electrical equipment through which an electric current of a particular voltage flows or may flow must be marked with signs indicating its type of protection and/or appropriate working voltage, their serial number and the date of manufacture. These markings must also appear on their packaging. A space must be provided on the outside of such PPE where the date on which the PPE began to be used can be indicated and where records showing periodic checks can also be indicated. The instructions for the use of PPE provided by the manufacturer must indicate the intended use and the nature and frequency of the dielectric tests carried out during the life of these PPE.

Reduction of electrostatic electricity can be provided in different textile production processes, such as yarns, fabrics, knitted fabrics, etc., depending on product type, the purpose of use and environment. Fabrics from natural fibres, such as cotton, can be sufficiently conductive for rapid dissipation of static charges, but these properties depend on moisture absorption from the environment. Synthetic fibres with low specific electrical conductivity are often used in the production of protective clothing, but such garments are able to accumulate an electric charge (Pinar and Michalak, 2006). Moisture absorption from the environment can also be used to improve the conductivity of chemical fibres, such as polyamide or polyester, which would otherwise be quite good insulators (Nurmi et al., 2007).

The fabrics used to control static electricity must meet at least one of the two basic requirements: they must not be easily electrified or, if electrically powered, must safely dissipate the charges faster than the charges accumulate. The main purpose of protective clothing in the electronics industry is to protect sensitive devices in the production phase from static electricity. Electrostatic discharge occurs in clothing on which static charge accumulates and electrifies. The functions of protective clothing are greatly reduced or even lost if the clothing is ungrounded. Grounding usually occurs through a direct collar or sleeve contact with a grounded operator. Improper use of the clothing can easily break the grounding path. Thus, alternative grounding methods should be considered or a permanent grounding connection control system should be designed to provide the desired function of protective clothing. Grounding of the electrostatic dissipating (surface conduction) clothing may take place through a conductive cuff flap directly in contact with the grounded body skin, through a separate grounded cord with a switch and through a connection with, for example, a conductive chair, grounded through a floor (man sitting) or conductive footwear through conductive floor (man standing) (Nurmi et. al., 2007; Gonzalez, 2005; Ono et al., 2003; Paasi et al., 2005).

An anti-static fabric (fabric with conductive yarns or coated with a conductive polymer coating) used in the manufacture of anti-static clothing must be certified. However, this is not enough because the garment must also meet the requirements of a special standard. The claim that the used garment indeed protects against static charge sparks is only acceptable if the design of the garment itself is approved by a certificate issued by the appropriate notified body. Anti-static clothing will not provide complete protection if extra clothes are worn on top: jackets, vests, etc. What should be worn under anti-static clothing is not so important. The basic requirement is that the anti-static clothing should be in contact with the wearer's skin at the wrist and neck area. All PPE, including protective clothing, will provide full protection if the instructions for use are followed (i.e. wear and maintenance instructions) provided that other PPE—respirators, shields, gloves, helmets, footwear—

are worn at the same time. For example, for anti-static gloves (as well as for anti-static footwear), the value of the electrical resistance must be less than 10^8 Ω. In this case, no static charge is generated, and no dynamic electrification process occurs. When working in an explosive, flammable environment, protection against electrostatic discharge and flame is necessary. This means that protective clothing should be anti-static and non-flammable. Intelligent textile production technologies have created fabrics that have both these properties at the same time. Simply put, the garment model of these complex fabrics should be such that it meets the requirements of anti-static and non-flammability simultaneously.

To protect against electrostatic discharge, the clothing must be grounded by direct contact with the wearer's skin or alternatively electrically connected through the wrist strap (Nurmi et al., 2007). Protective clothing that passes electrostatic charge must consist of a two- or one-piece suit and always cover the body, hands and feet. Clothing should be designed so that the charge dissipates from throughout the clothing and is in direct contact between conductive parts of clothing and human skin, such as the neck or wrists. The edges of the clothing, such as cuffs, trousers or collars, must improve the contact of the electrostatic fabric with the skin (EN 1149-1: 2000). Grounding is unnecessary if the clothing is made of low-electrification textile fabrics. It is very important that the sleeves of the ESD protective clothing cover the end of the garment's inner sleeve (Nurmi et al., 2007).

The purpose of the development of conductive fabrics is to create a balance between charge accumulation and dissipation. This is achieved by incorporating various conductive elements, such as chemical impurities, fibres or threads into the fabric structures. Modern electrostatic discharge protection fabrics today are heterogeneous composite materials, where conductive yarns are inserted into the insulating matrix of cotton, polyester, polyamide fibres/filaments in the form of webs or strips. Conductive fibres/filaments are increasingly produced as composites, such as matrices for conductive and insulating fibres/filaments (core conductive yarns, sandwich fibres, etc.) (Nurmi et al., 2007; Kim et al., 2004).

8. REQUIREMENTS FOR PROTECTIVE CLOTHING WITH ELECTROSTATIC PROPERTIES

The use of standards is voluntary in Europe unless they are made mandatory by law, such as the EU (European) Directives or Regulations. The valid standards for electrostatic discharge control can be divided into two groups (Von Pidoll, 2009):

- Avoiding electrostatic incendiary risk.
- Avoiding electrostatic damage to ESD-sensitive devices.

Both groups can be further grouped into three sections:

- Standards for basic requirements.
- Requirements for Special Purpose Products and Processes.
- Test methods.

These sections may be marked (Von Pidoll, 2009):

Standards: A set of requirements to be met by fabrics, products, systems or processes. They also describe procedures for determining parameters.

Technical reports: A set of technical data, test results, or knowledge printed as reference information for a particular fabric, product, system or process.

Standard test methods: Describes the measurement and evaluation procedure for one or more fabrics, products, systems or process properties and characteristics that give repeated test results.

Application of the standard: Procedures describing how to perform one or more operations or functions that may or may not be provided by a test result.

The textiles standards and personal protective clothing against unwanted dissipation of charges currently valid in Europe are described in Table 8.1.

Table 8.1. Standards for textiles and personal protective clothing against unwanted dissipation of charges

Test method	Title	Scope
EN 1149-5	Protective clothing. Electrostatic properties. Part 5 Requirements for material performance and model.	Requirements for protective clothing
EN 1149-1	Protective clothing – Electrostatic properties. Part 1: Test method for measurement of surface resistivity	Method for determining the electrostatic properties of protective clothing
EN 1149-3	Protective clothing. Electrostatic properties. Part 3: Test methods for measurement of charge decay	Method for determining the electrostatic properties of protective clothing
EN 1149-2	Protective clothing. Electrostatic properties. Part 2: Test method for measurement of the electrical resistance through a material (vertical resistance)	Method for determining the electrostatic properties of protective clothing

At present, the European Standard EN 1149-5 for performance and model requirements for electrostatic fabrics is widely used. This standard specifies that the electrostatic charge-dissipating fabrics used for protective clothing against incendiary discharge shall meet the following requirements:

- t_{50}<4s or S>0.2, performing tests according to EN 1149-3 standard, 2nd method, or
- surface resistance should be less than 2.5×10^9 Ω, for at least one surface, tested according to EN 1149-1 standard.

9. PROTECTION AGAINST ELECTROMAGNETIC PHENOMENA

9.1. Electromagnetic compatibility

Many sources of electromagnetic radiation (EMR) in our daily environment emit electromagnetic waves that can be the cause of interference in electrical and electronic devices. The nature of these sources varies widely: lightning, electric motors, digital computers, mobile phones. The effects of electromagnetic radiation on human health have not been definitively investigated, but many scientists state that EMR (the term EMR includes RFR, i.e., radio frequency radiation) have a detrimental effect on the human body, especially tissues, and hearing, visual organs, and cause cancer (Nurmi et al., 2007; Hoback and Reilly, 1988; Das et al., 2002; De Seze et al., 2001; Lai et al., 2007; Roh et al., 2008).

Increased electromagnetic radiation in an environment radiated by mobile phones, telecommunication systems, microwave ovens, machine-controlled radars and ultra-fast personal computers are challenging when trying to create shields protecting against such radiation. Protective shields for clothing should be aesthetic, flexible, comply with air permeability requirements and have a low surface density (Michalak et al., 2009). The harmful effects on human health from electromagnetic waves are shown in Figure 9.1.1.

Electromagnetic compatibility (EMC) is the ability of these devices to function in an environment without interference and radiation of electromagnetic fields (Nurmi et al., 2007).

Electromagnetic radiation must be reduced as soon as possible. This requires a suitable fabric, e.g. conductive textiles, which can act as protection against electromagnetic waves (Nurmi et al., 2007).

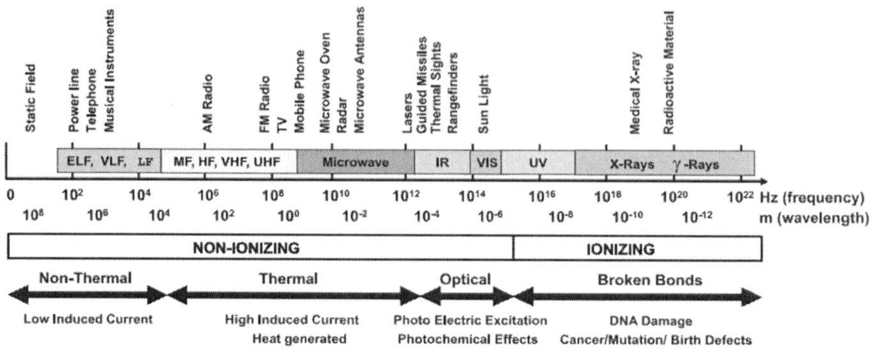

Figure 9.1.1. Graphic presentation of electromagnetic waves spectrum, application and effects (Roh et al., 2008)

Shielding is a very popular method of ensuring electromagnetic compatibility between electronics and electrical devices, and protection of people against electromagnetic emissions. Shields are used to isolate space (room, equipment, electrical circuit, etc.) from external sources of electromagnetic energy, or to prevent unwanted radiation from internal sources of electromagnetic energy. Most often, such shields are made of rigid metal material with well-known electromagnetic properties (Roh et al., 2008). An ideal barrier against electromagnetic radiation is a grounded, thick-walled, electrically conductive metal container with good magnetic conductivity (to minimize wall thickness, it should ensure that electrical and magnetic fields do not penetrate into the container). However, these barriers do not ensure the air permeability needed for the working staff, so their suitability is limited to protective devices. It is not possible to protect from the entire spectrum of electromagnetic radiation, including low-frequency fields, microwaves, infrared, visible, ultraviolet, X and gamma rays. Shields are designed for a specific, rather narrow frequency range (Michalak et al., 2006).

In recent years, much attention has been paid to electromagnetic wave barriers made of textile materials. These fabrics have excellent flexibility, lightness, some mechanical (fatigue, creasing) and chemical (corrosion and oxidation resistance) occurrence (Cheng et al., 2000; Cheng et al., 2001; Chen et al., 2008).

9.2. The main phenomena of electromagnetism

In the case of flat wave propagation, which can be assessed as the simplest type of propagation, the electric field and the magnetic field are built up of infinite parallel planes perpendicular to the direction of propagation (Nurmi et al., 2007).

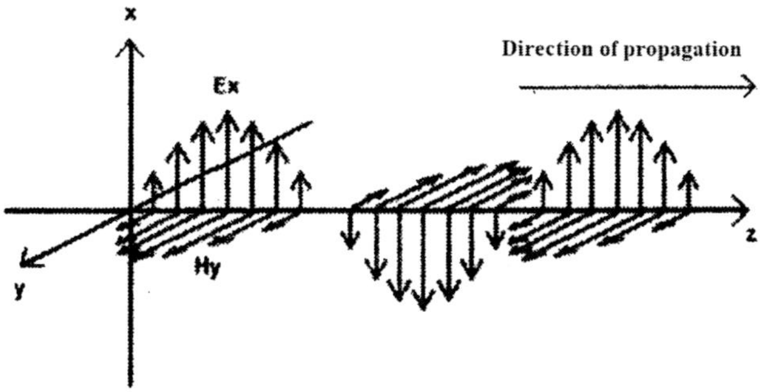

Figure 9.2.1. Instantaneous electromagnetic wave propagation diagram (Nurmi et al., 2007).

This can be achieved by using an electromagnetic source at larger distance r compared with the wavelength of the electromagnetic waves (far field region). The region of the flat waves increases significantly and behaves completely continuously, when this source is distant, meaning E (electric field) and H (magnetic field) are independent of the position in each plane (Nurmi et al., 2007).

In fact, if the electromagnetic wave comes from the sinusoidal source with a phase angle ω, then the electric field component in the x direction can be expressed using Maxwell's formula (Nurmi et al., 2007; Singh et al., 2012):

$$E_x = E_{x0} e^{-\gamma z} \tag{18}$$

A propagation constant γ appears in this expression, which depends on the wave phase angle ω and the speed of light in the vacuum c (Nurmi et al., 2007):

$$\gamma = j\frac{\omega}{c} \tag{19}$$

This dependency is only valid for lossless media; the wave does not suffer any attenuation when it propagates through the medium (Nurmi et al., 2007).

The magnitude of field strength is independent of z whereas the phase angle depends on this geometrical variable. By using Maxell's equation we can state similar relations for magnetic field $\vec{H}$. The relation between E_x and H_y is of a simple form (Saini and Arora, 2012):

$$\frac{E_x}{H_y} = \sqrt{\frac{\mu_0}{\varepsilon_0}} = Z_w \qquad (20)$$

where: μ_0 and ε_0– the absolute magnetic permeability and the absolute electric permittivity, respectively; Z_w– the wave impedance. This impedance may be connected with the electromagnetic power radiated from the RF source and its numerical value for lossless media is (Nurmi et al., 2007).

$$Z_w = 120\pi = 377\Omega \qquad (21)$$

The dielectric parameters – permittivity and permeability, of material are required in order to describe how electromagnetic waves interact with matter. From the material side, having the ability to make materials with specific complex permittivity (ε_w) and permeability (μ_w) over a suitable frequency range is essential. If the conductive textile material has non-ferrous nature, only the complex permittivity could be evaluated (Håkansson et al., 2007).

According to the distance r between the radiating source and the observation point, an electromagnetic irradiative region can be divided into three parts (see Figure 9.2.2) relative total wavelength λ of the electromagnetic wave. The region within the distance $r<\lambda/2\pi$ is the near-field while the distance $r>\lambda/2\pi$ is the far field. Between the two regions, as the distance $r\approx\lambda/2\pi$, is the transition region. For designing material for specific shielding application, it is imperative to possess in-depth data of both intrinsic and extrinsic parameters on which shielding effectiveness rely, together with appropriate theoretical relations correlating them with reflection, absorption and multiple-reflection loss components (Saini and Arora, 2012).

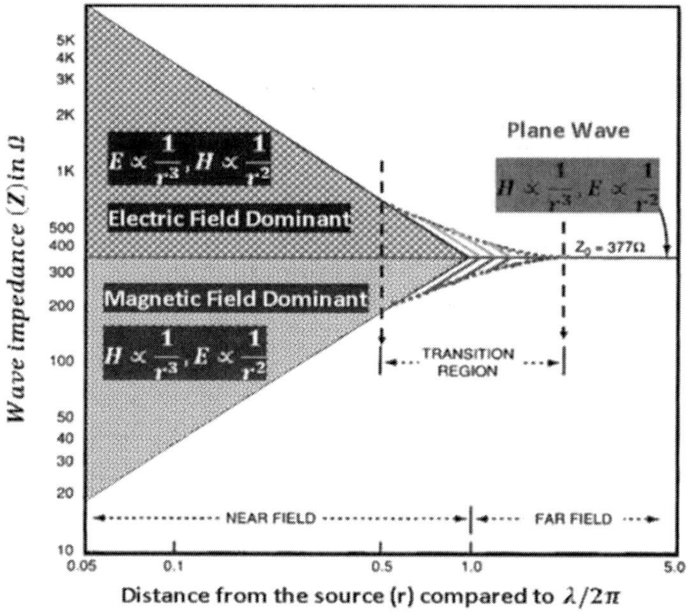

Figure 9.2.2. Dependence of wave impedance on distance from source normalized to λ/2π (Saini and Arora, 2012)

9.3. Electromagnetic shielding mechanism

Shielding fabrics act as barriers protecting against electromagnetic fields created in a very wide frequency range, which can be 50 Hz (60 Hz) for AC power supply, 100 kHz–30 MHz for medium and short wave radio broadcast stations, 100–500 MHz for FM and TV transmitters, 900 MHz and several GHz for mobile phones and radar sources. The metal shield that fully covers electronic devices can be the easiest way to realize an electromagnetic screen because a conductive electromagnetic shield weakens electromagnetic waves from external RF sources, when waves pass through the shield (see Figure 9.3.1) (Nurmi et al., 2007).

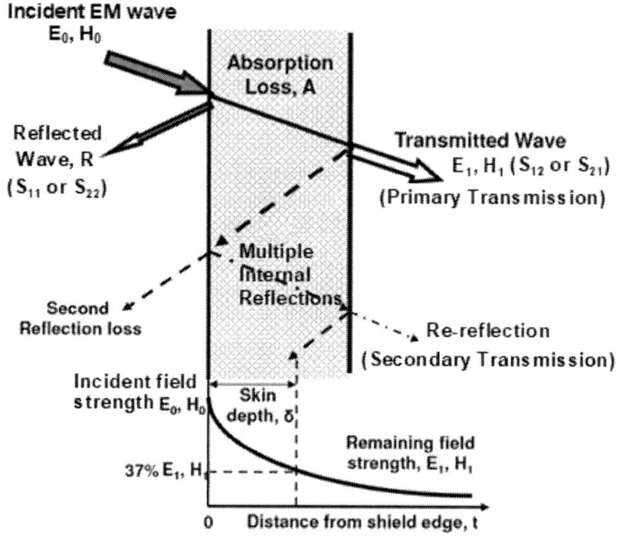

Figure 9.3.1. The multiple reflection phenomena on a thin screen

Waves passing through a barrier protecting against electromagnetic fields are affected by many physical mechanisms while interacting with molecules of the medium or object and weaken those fields. This interaction can be divided into two major steps: i) attenuation due to reflection and (ii) absorption attenuation (Maity and Chatterjee, 2018). The first is the wave reflection that occurs in the area between the free space and the barrier; in fact, part of the fallen wave is reflected from the surface until the other part of the wave penetrates through the material and becomes weakened when passing to the other side of the barrier (absorption loss) (see Figure 9.3.1). The absorption loss for an incident flat wave, penetrating through a good conductive medium (a good conductor must satisfy the condition: $\sigma >> \omega\varepsilon$) perpendicular to its surface, can be characterized by the depth of penetration so-called "skin depth" δ, and expressed by the formula (Nurmi et al., 2007; Maity and Chatterjee, 2018; Saini and Arora, 2012; Marchini, 1991):

$$\delta = \sqrt{\frac{1}{\pi \mu \sigma f}} \tag{22}$$

That means that over a distance of δ amplitude of the wave will be reduced by 30 % at a frequency f. Therefore if the thickness of the medium is much greater than the skin depth the wave amplitude that travels through this material is reduced dramatically (Maity and Chatterjee, 2018; Saini and Arora, 2012).

The absorption loss of one skin depth in a shield is approximately 9 dB. Skin effect is especially important at low frequencies, where the fields experienced are more likely to be predominantly magnetic with lower wave impedance than 377 Ω. From the absorption loss point of view, a good material for a shield will have high conductivity and high permeability along with sufficient thickness to achieve the required number of skin depths at the lowest frequency of concern (Singh et al., 2012).

It can be seen that multiple reflections and wave transmission occur on both sides of the material. This material is the cause of multiple wave motion, thus weakening more electromagnetic fields. The total field, passed through the barrier, is equal to the sum of all waves that passed multi-reflections (Nurmi et al., 2007; Avloni et al., 2007; Marchini, 1991; Lai et al., 2007; Cheng et al., 2000).

The time changing magnetic field induces currents on the enclosure, these induce a magnetic field inside the enclosure with opposite phase angle, so the resulting magnetic field decreases inside when the frequency increases.

Thus, shielding efficiency (*SE*) can be divided into three phenomena: reflection loss (*R*), absorption loss (*A*), and multiple reflections (R_{multi}) (Maity and Chatterjee, 2018; Saini and Arora, 2012; Yuping et al., 2006; Cheng et al., 2000; Chen et al., 2008):

$$SE = \mathrm{R} + \mathrm{A} + \mathrm{R}_{\mathrm{multi}} \tag{23}$$

9.4. Shielding efficiency

The control of electromagnetic interference (EMI) shielding efficiency (SE) depends on the conductivity of the fabric. Fabrics with a specific surface resistivity of less than 10 Ω are suitable for protection against electromagnetic interference and fabrics with larger specific surface resistivity are suitable for anti-static purposes (Geetha et al., 2005).

The effectiveness of electromagnetic barriers is usually characterized by shielding efficiency (SE), expressed as the ratio of the electrical (or magnetic) field strength without the barrier and the electric (or magnetic) field strength inside the barrier. By this definition, we can express the electric field shielding efficiency (dB) as (Nurmi et al., 2007; Avloni et al., 2007; Yuping et al., 2006; Das et al., 2002; Geetha et al., 2005; Maity and Chatterjee, 2018):

$$S_E = 20\lg\left|\frac{E_i}{E_t}\right| \tag{24}$$

The electromagnetic efficiency of the magnetic field is:

$$S_H = 20\lg\left|\frac{H_i}{H_t}\right| \tag{25}$$

where: E_i– incident electrical field strength; E_t– transmitted electrical field strength; H_i– incident magnetic field strength; H_t– transmitted magnetic field strength.

For a single layer, the theoretical EMI SE (dB) can be written as (Kim et al., 2003):

$$EMI\ SE = 20log\left(1 + \frac{1}{2}\sigma d Z_0\right) \tag{26}$$

where: σ– conductivity; d– the thickness of the sample; Z_0– the free space wave impedance, 377 **Ω**.

In designing and selecting coated or multilayer textile materials for the construction of shields, masking elements or special protective clothing, including camouflage clothes, a significant element is knowledge of the following properties of these materials within the required band of frequency (Brzeziński et al., 2009):

- Reflection coefficient– the reflection of the electromagnetic wave from the media boundary resulting from the misfit of the wave impedance of the media.
- Absorption coefficient– the absorption of EMR wave energy by the material (including the phenomenon of internal reflections).
- Transmission coefficient– EMR penetration into the medium behind a barrier (shields or clothes etc.).

By measuring the reflectance (*R*) and the transmittance (*T*) of the material, the absorbance (*A*) can be calculated using the following equation (Abdelal, 2018):

$$A=1-T-R \tag{27}$$

where: *R* and *T* are the square of the ratio of reflected (E_r) and transmitted (E_t) electric fields to the incident electric field (E_i), respectively, as follows (Maity and Chatterjee, 2018; Maity et al., 2013):

$$R = \left|\frac{E_r}{E_i}\right|^2 = |S_{11}(or\ S_{22})|^2 \tag{28}$$

$$T = \left|\frac{E_t}{E_i}\right|^2 = |S_{21}(or\ S_{12})|^2 \tag{29}$$

In a two-port network analysis system, the scattering parameters *S*, i. e. S_{11} (S_{22}), S_{12} (S_{21}) can be correlated to the reflection and transmission coefficient (Nayak et al., 2013).

Nicolson-Ross-Weir technique showed that for electrically thin materials the scattering parameters S_{11} and S_{21} can be described as (Saini and Arora, 2012; Abdelal, 2018):

$$S_{11} = \frac{\mathrm{R}(1-\mathrm{T}^2)}{(1-\mathrm{R}^2\mathrm{T}^2)} \tag{30}$$

$$S_{21} = \frac{\mathrm{T}(1-\mathrm{R}^2)}{(1-\mathrm{R}^2\mathrm{T}^2)} \tag{31}$$

If in an EMI shielding material, the effect of multiple reflections is considered as negligible, then the relative intensity of the effective incident wave inside the material after reflection is based on the quantity *(1-R)*. So, the effective absorption coefficient can be described as (Nayak et al., 2013):

$$A_{eff} = \frac{(1-\mathrm{R}-\mathrm{T})}{2a(1-R)} \tag{32}$$

where: SE_R and SE_A of the shielding material are correlated with the reflection (*R*) and transmission (*T*) coefficient by the following equations (Nayak et al., 2013; Abdelal, 2018):

$$SE_R = -10\log(1-R) \tag{33}$$

$$SE_A = -10\log\left(1-A_{eff}\right) = -log\frac{T}{1-R} \tag{34}$$

where: R– reflection; T– transmission; A– absorption coefficient of shielding material.

With reference to the theory presented and obtained expressions (Rubežienė et al., 2018), reflectance (R) and transmittance (T) can be expressed as:

$$R = \left[\frac{\eta\sigma_S}{2+\eta\sigma_S}\right]^2 \quad (35)$$

$$T = \frac{4}{(2+\eta\sigma_S)^2} \quad (36)$$

where: η– the impedance of free space; σ– the surface conductivity.

Absorption (A) in the layer can be easily calculated multiplying the drop of the amplitude of the magnetic field with the amplitude of electric field in the layer leading to:

$$A = \frac{4\eta\sigma_S}{(2+\eta\sigma_S)^2} \quad (36)$$

The typical behaviours of S_E and S_H versus the frequency are given in Figure 9.4.1. It is clearly seen from Figure 9.4.1, that above f_t, the shielding effectiveness of the electric field increases to join the shielding effectiveness for the magnetic field at higher frequencies. This characteristic frequency f_t corresponds to the condition where the skin depth δ becomes smaller than the shield thickness t:

$$\delta \leq t \text{ then } f_t = \frac{1}{\pi\mu\sigma t^2} \quad (37)$$

This phenomenon appears when the enclosure dimension becomes oversized compared to the wavelength.

We can notice that apertures in enclosures reduce dramatically the shielding effectiveness. Indeed, these apertures will change current and charges distributions at the enclosure surface area producing leakage of the magnetic and electric field inside. The aperture size and contact resistance of yarns in fabrics are critical parameters in determining shielding effectiveness (Shateri-Khalilabad and Yazdanshenas, 2013).

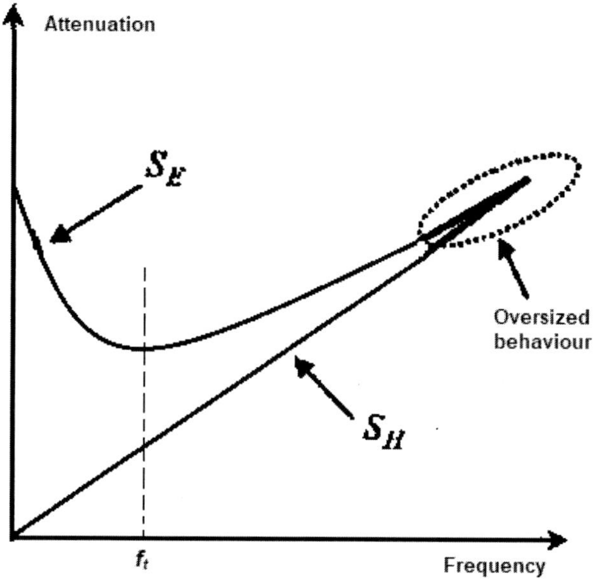

Figure 9.4.1. Effectiveness of S_E and S_H in frequency (Nurmi et al., 2007)

The decrease in electromagnetic energy is characterized not only by the size of the *SE* but also by the size of the *IL*. *IL* is the magnitude of the attenuation of the signal passed through the material in the measured channel (Avloni et al., 2007; Avloni et al., 2008; Maity and Chatterjee, 2018):

$$IL_{db} = 10\lg\left(\frac{U_0}{U_1}\right) \tag{38}$$

where: U_0– channel output voltage (power) without the test sample; and U_1– the same voltage (power) with the test sample.

The holes in the barrier significantly reduce the effectiveness of shielding. These holes replace the current, so the charge distribution on the surface of the sample causes the magnetic and electrical fields to leak inside the barrier (Nurmi et al., 2007).

There is no doubt that the best EMI shielding and absorption material should have high conductivity and high magnetic permeability. Therefore, devices made of metal impurities are considered the best shielding devices. Materials used for radio frequency (RF) shielding are mainly copper, aluminium or silver conductive paints. Such barriers reflect waves as these metals have high electrical conductivity. Magnetic field shielding materials must have magnetic penetration to absorb fields. If RF barriers have a low amount of metal impurities or do not have them at all, they become barriers to the broadband radio system that prevents EMI and RF interference (Avloni et al., 2008).

Avloni et al. (2007), using the TEM measuring chamber, assessed the decrease in the electromagnetic energy of metallized and polypyrrole-coated woven and non-woven fabrics in *SE* and *IL* values. They found that EMI shielding efficiency increases with increasing electrical conductivity of the material surface, which increases the reflectance of electromagnetic radiation. The high reflectivity coefficient is due to the high conductivity of the surface coating of the composite. The shielding performance of polypyrrole-coated materials can be controlled by changing the electrical conductivity of the coating. The authors have also investigated that polypropylene non-woven materials have a high absorption coefficient compared to metallized materials. The shielding efficiency of metallized textiles is usually dependent on the energy of reflection rather than the energy of absorption. Polypropylene materials have good absorption properties and provide good absorption of electromagnetic radiation.

Cheng et al. have studied the screening efficiency of stainless steel/polyester/glass/polypropylene knitted fabric composites (Cheng et al., 2006). They note that the effectiveness of electromagnetic shielding can be linked to many parameters of the material structure, e. g. the density of the material, the number of layers and the amount of conductive yarn. The quantity of stainless steel in the composite laminate has a strong impact on EMS (electromagnetic shielding) performance characteristics. The more metal threads present in the material and the higher the density of the knitted material, the higher the EMS efficiency of the composite. Similar research results were described by Cheng et al., (2001; 2006). Experiment researchers have confirmed that EMS performance depends on the amount of conductive yarn and fabric density: more yarns results in better shielding. It has also been found that the conductivity of the conductive fabric with the smallest pores is the greatest. Similar findings were made by Lai et al. (2007) studying polyester materials coated with silver, copper, aluminium and titanium.

De Temmerman (1992) studied non-woven polyamide materials, coated with various concentrations of the metallized coating. The author has found that high shielding efficiency is achieved by increasing the metal concentration in the coating. He also noticed that the shielding level decreases with increasing fabric specific resistivity.

Non-woven materials with flax and polypropylene for electromagnetic radiation shields were investigated by Michalak et al. (2006) and Brazis et al. (2000). Thermal resistance and electrical resistance were measured in dry conditions. Researchers found that the amount of flax significantly reduces the electrical resistance of non-woven materials. The biggest resistance has pure polypropylene non-woven materials, and the resistance of pure flax is about four rows smaller.

For EMI shielding applications, typically an SE of at least 20 dB (indicates that 99 % of the electromagnetic energy is reflected or absorbed by the material) is needed. An SE of 30 dB indicates that 99.9 % of the EM energy is reflected or absorbed by the material, with only 0.1 % exiting the shielding material (King et al., 2015).

High material dielectric constant values and high conductivity provide materials with shielding efficiency at high-frequency electromagnetic fields. However, at low frequencies, the attenuation of the magnetic field H is very difficult. Materials used at such frequencies must have high magnetic permeability (Nurmi et al., 2007). This is done by using ferromagnetic materials. The construction of textile materials, as well as the fibre composition, can have a significant impact on the properties of electrical materials (Ouyang and Chappell, 2005). Ferromagnetic textile materials can be produced by applying suitable materials on the surface or by incorporating them into fibre polymers (Koprowska et al., 2004).

Recently, conductive polymers have become good candidates for low-frequency shielding, thanks to their intrinsic properties, especially polyaniline (Nurmi et al., 2007).

9.5. Evaluation of electromagnetic shielding properties for textiles

Shielding material is typically used to encase an electronic product to prevent the enclosed product from emitting electromagnetic or radio frequency (RF) energy. The shielding material either absorbs or reflects the energy inside the material. A shield can be characterized by its shielding effectiveness, which is dependent on the material of which the shield is made, the thickness of the shield, the frequency, the distance from the source to the shield and the quantity and shape of any shield discontinuities. The attenuation provided by a shield results from the following three mechanisms (Perumalraj et al., 2009):

1) Incident energy that is reflected by the surface of the shield because of the impedance discontinuity of the air-metal boundary. This mechanism does not require a particular material thickness, but simply an impedance discontinuity.
2) The energy that does cross the shield surface i.e. is not reflected, it is attenuated in passing through the shield.
3) The energy that reaches the opposite face of the shield encounters another air-metal boundary and thus some of it is reflected back into the shield.

The electromagnetic shielding effectiveness (SE) of the element is characterized by its electric conductivity, permittivity, and permeability, parameters of source and properties of the ambient surrounding. Basic proposed numerical models of fabrics shielding efficiency are based either on electrical property (especially volume conductivity) of element (Perumalraj et al., 2009; Keith et al., 2005; White and Mardiguian, 1988; Colaneri and Schacklette, 1992) or on analysis of leakage through the openings in textile (Šafářová et al., 2015). A quick method to determine the suitability of conductive material is to measure vertical resistivity. If materials have a vertical resistivity above 100 Ωcm, they are no longer good as an EMI shielding material but can be used for electrostatic dissipation (ESD) application. Vertical resistivity measurements sometimes suffer from inhomogeneous filler distribution, which creates leakage but still produces a good macroscopic conductivity. A qualified shielding product may fail in the vertical resistivity measurement due to the depletion of fillers in the skin layer (Roh et al., 2008). The electrical resistance of EMI fabric changes according to yarn to the yarn of fibre to fibre contact (Neha et al., 2013). The surface resistance of EMI fabric is dependent on the exposure of conductive material (core conductive yarn) (Mehmood et al., 2012).

Determining the level of attenuation of an EMI shield can be complex and the methods often vary according to the particular shield application. However, a variety of standards are adopted to measure the performance of planar shielding structure (Celozzi et al., 2008). The more common techniques for testing shielding strength for plane materials (textiles) may be classified into two main groups: (i) measurement methods for Near-Field and (ii) measurement methods for Far-fields. The first group includes dual chamber and dual TEM cell methods; the second group includes coaxial transmission line, KEC, waveguide, space free methods. In general, the most popular methods are Field Test, Coaxial Transmission Line Test, Shielded Box Test, and Shielded Room Test (Geetha et al., 2009; Cheng, 2000; Roh et al., 2008).

9.5.1 Open field or free space method and time domain approach

The free space method is suitable for flexible thin samples which are difficult to measure using conventional microwave measurement techniques, such as waveguide methods, dielectric probes and coaxial transmission lines (Rupprecht, 1999). Free space methods using larger samples avoid many of the sample size and contact problems associated with waveguide measurements (Amiet, 2003). This method is highly reproducible and a very large number of measurements are done on each sample, and thus provide statistically reliable results. The only drawback of the method is the requirement of sophisticated and expensive equipment (Rupprecht, 1999).

The open field or free space method is used to evaluate the practical shielding effectiveness of a complete electronic assembly. Thus, this test measures the radiated emissions that escape from a finished product. The open field test method requires a large open field, because the method involves mounting the device at a distance of 30 m from a receiving antenna and recording the radiated emissions and there should not be any metallic or conductive object between the sample and the receiving antenna (Fu et al., 2017; Geetha et al., 2009; Rubežienė et al., 2015).

From the measurements, which will be given as graphs, it is seen that open area measurements can be affected by environmental conditions. More peaks in the graphs and changes of shielding effectiveness raised by reflections are observed through the whole frequency band. The conducted emissions transmitted down the power line are also recorded during the open field test as well. Average shielding levels are low compared to chamber measurements. (SEÇKİN UĞURLU et al., 2015; Geetha et al., 2009).

The use of a time domain source makes it possible to differentiate the direct path (through the test sample) and the indirect path (diffracted). The signal is reordered at the receiving end. An impulse generator is used to generate an impulse through a TEM (Transverse Electromagnetic) horn antenna. Insertion loss measurements were found by measuring the difference of power recorded at the transmitting and receiving ends (Bagavathi, 2015).

The measurement procedure includes the following two steps (Bagavathi, 2015):

1) calibration-measurement of the transmitted signal without material under test (*P0*);
2) measurements of the transmitted signal as a function of incident angle in the presence of material under test (*P1*);
3) the ratio *P1/P0* is the power transmittance needed to extract the complex permittivity.

9.5.2. Shielded Box method and waveguide method

The shielded box method is widely used for comparative measurements of test specimens of different shield materials. The test comprises of a metal box and an electrically tight seam that has a sample port in one wall and is fitted with a receiving antenna. A transmitting antenna is placed outside the box and the intensity of signals received by the antenna is recorded both through the open port and with a test specimen fitted over the port (Geetha et al., 2009). Effectiveness of the shielding can be determined by the ratio of electric power transmitted by the test material and the reference material in attenuation measurements for shielded enclosures within the frequency range of 100 kHz to 10 GHz (Tong, 2016).

The measurement range in this method is divided into 3 sub-ranges (Tao et al., 2016):

- Low range – from 9 kHz (50 Hz) to 20 MHz – for magnetic component (*H*)
- Resonant range – from 20 MHz to 300 MHz – for the electrical component (*E*)
- High range – from 300 MHz to 18 GHz (100 GHz) – for plane wave power (*P*)

The drawback of this method is that adequate electrical contact between test specimens and the shielded box is difficult to achieve. The results from different laboratories show poor correlation (Geetha, 2009). The shielded box test is the most reliable for materials in near-field conditions i.e., up to 500 MHz. Furthermore, inadequate contact between the sample and the shielded box may affect the repeatability and reliability of results (Fu et al., 2017).

9.5.3. Shielded room method

The shielded room method has been developed to overcome the limitations of the shielded box method. The general principle of this technique is the same as for the shielded box method, except that every element of the measuring device, signal generator, transmitting and receiving antennas and a recorder is isolated in separate rooms to eliminate the possibility of interference. In addition, the antennas are located in room-sized anechoic chambers and the test specimen size is greatly increased, typically of the order of 2.5 m^2 in area. The frequency range over which reliable results can be obtained is greatly extended and the reproducibility of the data is significantly improved (Geetha et al., 2009; Fu et al., 2017). It is possible to obtain repeatable results by using this method, especially in externally disturbing conditions (Fu et al., 2017).

9.5.4. Coaxial transmission line method and TEM cell method

The coaxial transmission line method is the most common one due to its suitability to measure small-sized, flat, and thin conductive samples in an extended range of frequency (Fu et al., 2017). This method for the measurement of shielding effectiveness overcomes the limitations of the shielded box technique. The results obtained in different laboratories are comparable. In addition, the coaxial transmission line can also be used to resolve the data into the reflected, absorbed and transmitted components. Tests are carried out on small doughnut-shaped samples. The specimen must completely fill the waveguide cross section or large errors result. The sources of errors in the measurement are due to operator errors, specimen-caused errors, and measurement system errors as well (Bagavathi, 2015). This method has the advantage that thickness measurements are not required, which is usually a major source of inaccuracy when the sample is not exactly flat (Amiet, 2003). The measurements can be made at specific frequencies, mostly 0.01 MHz to 1000 MHz, using a modulated signal generator, crystal detector and tuned amplifier or alternatively, in a swept mode using a tracking generator and spectrum analyzer as a receiver (Geetha et al., 2009). This measurement unit works on the principle of capacity (Tao et al. 2016). In the point-by-point mode, the system is first set up at a given frequency without the specimen holder in the line. The variable attenuator is set to maximum and the signal level is recorded. The specimen holder is then inserted into the line and the attenuator reduced until the same reading as before is recorded. The attenuation of the signal obtained is a direct measure of the shielding effectiveness of the specimen. To obtain the spectrum of responses the procedure is repeated at a series of different frequencies. Obviously, this point-by-point approach is time consuming. In the swept mode a tracking generator driven by a spectrum analyzer replaces the generator. The spectrum analyzer presents the response of the system as a single curve on a display screen in a few minutes. A dynamic range of about 80 dB can be obtained with standard coaxial cables (Geetha et al., 2009; Tao et al. 2016).

The dual TEM cell allows simultaneous measurements of electric and magnetic polarizabilities of test material fixed in a small aperture, which is not possible with a single TEM cell. The dual TEM cell consists of two TEM cells one of which is coupled via a small aperture in the shared conducting wall. The aperture transfers power from the driving cell to the receiving cell. The dual cell simulates both high and low impedance near-field at the same time. The effect of a dual cell during SE measurement could be removed by measuring insertion loss in an empty cell and subtracting it from the insertion loss measured with the cell loaded with sample (Bagavathi, 2015).

10. DEVELOPMENT OF CONDUCTIVE TEXTILES

10.1. Special fibres and filaments

Textile fibres are generally electrical isolators. The synthetic fibres used in the textile industry consist of polymers. As the chemical structure of the polymers does not store free electrons, there is an absence of electrical carriers, which would form the electrical current upon the application of an electrical potential. The typical specific resistivity of the polymer used in the production of synthetic textile fibres is higher than 10^{10} Ωm order of magnitude. After a chemical metallization procedure, the electrical conductivity of the man-made textile fibres increases steeply. The level of the electrical conduction of the textile fibres reaches higher values, so that they can be considered as electrical conductors suitable for technical applications. The electrical characteristics of the yarns do not seem to be directly proportional to the electrical characteristics of the fibres. Principally, the electrical conductivity of the fibres depends on the quantity of the metal deposited. Further, the electrical character of the yarns is made up of the electrical characteristics of the constituting fibres and of the structure of the yarn. The electrical conductivity of yarns is highly affected by the contact effects taking place between the fibres. The contact resistance is often more significant than the nominal resistance of the material. Consequently, the yarn structure affects essentially the electrical behaviour of the textile yarns made of electrically conductive fibres (Vassiliadis et. al., 2004). For example, for a mixed yarn consisting of 50 % natural fibres and 50 % stainless steel fibres, only the sum of the cross sections of the stainless steel fibres needs to be taken into account, thus determining the linear resistance of the yarn.

Many special fibres with good electrostatic properties have been developed over the last few decades. Many fibres are modified forms of standard fibres which, depending on their use, completely or partially replace standard fibres. Conductive properties of fibres, produced at low concentrations from standard fibres, are improved by the addition of moisture-absorbing impurities during production. In this way, their electrical activity becomes moisture-dependent and their conductivity is based on ion conductivity. The latest modified forms of standard fibres, such as bicomponent fibres, also known as chemical fibres, are mixed metal or carbon fibres. The electrical conductivity of bicomponent fibres is greater than that of the fibres with which they are mixed and the electrical activity of chemical structure fibres is independent of moisture (Nurmi et al., 2007).

Currently, studies and industry have produced conductive yarns out of whole metals known as metallic yarns, hybrid or composite yarns out of a combination of metallic yarns and conventional textile yarns, yarns laminated with conductive films, yarn coated with conductive solutions, and yarns made from nano-materials (Raji et al., 2017).

Some major types of conductive fibre production (Nurmi et al., 2007):

- Polymerization of polymers with moisture-absorbing reactants by polycondensation or coupling;
- The polymer is displaced with a moisture-absorbing polymer by forming a bicomponent fibre (shell-core, matrix-fibril);
- The polymer is displaced with a moisture-absorbing non-polymeric mixture.

Two main types of fibre structure are commonly used: surface conductivity and core conductivity (see Figure 10.1.1). Occasionally the term semi-conductive fibre is used, which means that in the fibre structure the conductive component is partially on the surface and partially embedded in the fibre (Nurmi et al., 2007; Kalliohaka et al., 2005; Paasi et al., 2004).

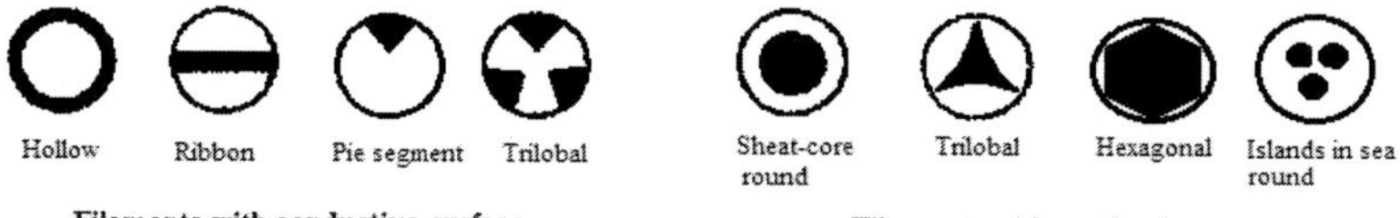

Figure 10.1.1. Examples of conductive filament diameter

10.2. Conductive yarns and fabrics

10.2.1. Metallized textiles

The first approach towards affecting the conductivity of the textile fabrics is the use of metallic wires and thin metallic tapes. The metallic wires are interlaced into the fabric structure and it gives the required electrical characteristics to it. The electrical conductivity of the fabric is controlled through the wire diameter and the density of the wires in the structure of the fabric. The method of wire weaving into fabrics obviously results in the production of fabrics characterized mainly by extremely limited flexibility, increased weight and problems related to the forming of the final product (Vassiliadis et al., 2004). Wires have been found to not provide a good comfort level (Rakshit and Hira, 2014). Most metallic wires are obtained by a drawing or bundle drawing process. The wire diameter can be reduced by several successive drawing steps. Because of this drawing process, a thermal treatment is required afterwards, in order to avoid the thread being too brittle. Through successive drawing steps, a minimal diameter of 10 μm can be obtained of copper wires, but usually they have a diameter of 40 μm) and 6 μm diameters can be obtained for stainless steel wires, but usually wires with a diameter of 12 μm are used.

Copper products: continuous fibres 100 % copper (tin-plated), or wrapped.

Metal yarns are a large group of different types of conductive fibres, where conductivity is provided by stainless steel wire, metal particles/impurities, metal oxides or metal salts. Stainless steel yarns are made exclusively of metal fibres (Nurmi et al., 2007).

Electrically conductive materials are applied in three ways:

- As pure material: 100 % fibres
- As a coating or plating, with or without adding an adhesive
- As a composite: polymer(s) mixed with electrically conductive particles.

Metals fibres can be incorporated in the textile yarns or fabric structures in various ways. Few of the methods are depicted below (Rakshit and Hira, 2014):

- Metal-wrapped yarns are a composite of metal and textile yarn. Conductive metallic yarn is wrapped with one or more strands of non-conductive textile filaments.
- Metal-filled yarns are obtained by having a fine metal wire as core covered with non-conductive fibres. The textile covering protects the core metal and helps it to withstand physical stress and provides insulation.
- Metal fibre does not form the core, rather it replaces one or all strands in the plied yarns.

Conductive filaments containing metal can be core conductive fibres, sheath-core filaments or surface conductive filaments. Metals can also be introduced into the polymer matrix comprising the synthetic textile material. The conductive element can be silver, nickel, copper, aluminium, cobalt or metal particles/impurities (Nurmi et al., 2007). Copper is the most conductive material, after gold and silver. Its electrical conductivity ranges around $6x10^7$ S/m or a density of 8.9 g/cm^3.

The electrically conductive staple fibres can be used for the production of electrically conductive yarns according to the traditional production of staple yarn. No special equipment is necessary. They can be blended with conventional fibres in order to combine the electrical characteristics with the physical characteristics of the conventional fibres. This technique needs the presence of very thin metallic fibres to be mixed with the typical synthetic fibres. Both the production of metallic fibres and their processing have imposed several difficulties. The usual spinning machines are unable to process metallic fibres since they are harder than the synthetic ones and they may damage the surface of the mechanical elements involved in the spinning procedure. The materials produced are of less flexible and heavier (Vassiliadis et al., 2004).

Conductivity can also be obtained by coating a metallic surface on a "traditional textile". The applied conductive materials are mainly silver, copper and copper sulphide.

Copper fibres can also be insulated by covering them with a plastic sheath. Because of the weak mechanical properties of copper fibres, they are not combined with (i. e. wrapped around) a traditional textile yarn that provides the mechanical support of the yarn for further processing. Copper yarns are only available as continuous filaments. Copper fibres have a low fatigue resistance (they break as a result of successive folding). They corrode easily in the presence of humidity and acid. This corrosion can be avoided by depositing (through electrodeposition) a thin layer of silver or tin, for example, on the surface.

Stainless steel fibres have a good corrosion resistance, which makes it possible to use them without adding a protective layer. They also have good mechanical properties and are resistant to folding and torsion. Pure stainless steel yarns can be woven, knitted or performed in a non-woven structure. These textile products are used in applications where flexibility and a stable product under varying thermal conditions are required, such as filtration at high temperature or protection in the glass industry.

Stainless steel fibres can be mixed with traditional textiles such as cotton or polyester. This results in a product that is less heavy, has a better touch, -lower electrical conductivity and a lower price.

Stainless steel is used in textile applications for several purposes:

- Anti-static purposes: to avoid deposition of dust or electrical discharge; the content of stainless steel fibres is about 0.5 to 1 % in volume
- Electromagnetic shielding: according to the principle of Faraday's cage; the content of stainless steel fibres is about 1 to 2 % volume
- Radar shielding for military purposes.

Stainless steel products: continuous fibres 100 % stainless steel; discontinuous fibres (long and short) closely blended, twisted, wrapped and so on.

Filaments containing conductive metal oxides can be surface conductive, sheath or sheath-core fibres, concentric core or eccentric core-shaped fibres. Filaments with conductive metal salts are produced by chemically forming salts on the surface layer of fibres (Nurmi et al., 2007). The conductivities of metal oxides, such as SnO_2, ZnO, Sb_2O_3, are not as high as those of metal materials, but they are still useful in making conductive or semi-conductive textiles for some specific applications (Zhang, 2011).

Metal coating on textile, i.e. the deposition of a metallic layer on a textile product (i.e. fibre, yarn, woven or knitted structure) is achieved by means of vacuum pulverization. Metal is evaporated in a chamber and condenses on the textile surface. This technique allows the deposition of a thin layer with controlled and regular thickness. Typically, layer thicknesses of 1 to 2 μm are obtained. The advantage of this technique is that the produced fibres have a density close to that of the original fibre and the characteristics of the textile are maintained. The electrical properties depend on the layer thickness and on the applied metal. Silver and copper sulphide are most frequently used.

The technique is expensive and not really applicable on large surfaces because the production process takes place in a vacuum and is often discontinuous. Furthermore, when using silver (even though not very oxidizable) as a conductive material, problems can arise with humidity, salty environments and oxygen because the surface oxidizes.

Copper sulphide deposits are less conductive. In general, the thickness of the conductive layer generates problems related to durability.

Metal coated textiles are used in the same applications as stainless steel fibres, i.e. anti-static applications, electromagnetic shielding, heating textiles (only for silver).

This metallization technique can be used with almost all metals. However, the best compromise between cost, conductivity and oxidation is reached with silver.

Although metals are the most common EMI shielding material, EMI shielding by absorption rather than reflection is of major interest. Metals could not be used as an absorbent since their shallow skin depth leads to shielding through a reflection mechanism. For high EM shielding effectiveness, the reflection mechanism is often adopted due to the presence of free electrons in the metal structure (Šafářová and Militký, 2017). Reflection loss for plane waves is greater at low frequencies and for high conductivity materials. Electromagnetic radiation at high frequencies only penetrates the near surface region of an electrical conductor. The electric field of a plane wave penetrating a conductor drops exponentially with increasing depth into the conductor. The reflection loss decreases with increasing frequency, whereas absorption loss increases with increasing frequency. Materials with high absorption loss and low reflection loss are highly effective in shielding electromagnetic energy (Cheng et al., 2001). Stainless steel has high absorption and low reflection of electromagnetic energy in low frequencies; it has high magnetic permeability (Šafářová and Militký, 2017; Cheng et al., 2001).

Chen et al. (2004) described the successfully developed rotary spinning method of stainless steel/polypropylene, copper/polypropylene and stainless steel/copper/polyamide yarns and the production of conductive composites from woven-knitted fabrics reinforced with polypropylene, manufactured using a new method. Solid copper, stainless steel wire and non-wovens are best knitted with the use of conductive hybrid technology. The EMSE, ESI and specific surface resistivity of woven-knitted reinforced composites are significantly influenced by the amount of copper and stainless steel, and the contact points between the conductive elements in the composite material. Resistance may vary by changing the structure of woven-knitted reinforced materials, the lamination angles, and the linear density of yarns used to form knitting loops and incorporation of the weft yarn. The properties of EMSE and ESI of conductive woven-knitted composites, reinforced with polypropylene, are very good.

Šafářová and Militký (2017) concluded, that electromagnetic shielding effectiveness increases with increasing metal fibre content and that electromagnetic shielding effectiveness will not increase dramatically above 2450 MHz. As the frequency increases, the wavelength of the electromagnetic wave decreases and becomes closer to the size of the fibre.

Redlich et al. (2014) determined the microwave parameters of textile structures developed (woven and knitted fabrics with incorporated steel fibres) using waveguide applicators and the "Time Domain" method. They concluded, that materials used to camouflage personnel and their equipment, should feature the lowest reflectance and transmission coefficients, as well as the greatest absorption factors. The analysis of test results indicated that woven fabric samples featured reflection coefficients that were too high and attenuation coefficients that were too low. Measurement of textiles developed indicated that an irregular distribution of conductive fibres within the structure has a positive effect on their anti-radar camouflage properties. The authors also concluded that multiple layers of such conductive textiles allow for complementarities between the camouflage parameters of each one.

Telipan et al. (2017) carried out shielding effectiveness measurements in an anechoic chamber with conductive textiles containing natural fibres (cotton and wool), synthetic fibres (polyester and nylon) with conductive yarns of metals (like copper and stainless steel) or carbon. The measurements were taken in the frequency range of 1–18 GHz. They noted that efficiency of shielding might be achieved using technical textiles only in high-frequency ranges (16–18 GHz). As an option to achieve the desired optimum shielding in the 1–18 GHz range, an additional fabric layer with shielding properties is proposed. They also concluded, that technical textiles that contain fibres with copper and stainless steel will present shielding properties better than fibres which contain carbon. This behaviour action of the external electromagnetic field is due to the diamagnetic properties of copper and stainless steel.

Duran and Kadoğlu (2015) investigated fabrics with silver containing core yarns and silver containing blended yarns. They determined that shielding effectiveness decreased with increasing frequency in investigated ranges, due to a smaller wavelength in higher frequencies, according to the formula $f=c/\lambda$, where f is the frequency, c is the speed of light and λ is the wavelength of the EM wave.

10.2.2. Textiles with carbon

Besides metals, carbon materials, such as carbon black, carbon nanofiber, carbon nanotubes, graphite and graphene, may be used as EMI shielding materials because of their high specific surface area and low density, versatile processability, and excellent electrical conductivity. Carbon is known for its excellent absorption characteristics. Nanoscale materials are reported to have the ability to fill up the vacancy of the conductive network formed by conductive materials of different shapes, resulting in a denser and more complete conductive network (Shahidi and Moazzenchi, 2018).

Many conductive filaments are made from conductive carbon by acetylation. Various types of conductive filaments are produced using carbon as impurities. The filament may consist of a high concentration of carbon, or the carbon may be inserted into the core of the shell-core bicomponent filament. The first products were manufactured by coating the filaments with a resin of high carbon content. Active carbon fabrics have very good conductivity and stability (Nurmi et al., 2007; Cisło et al., 2003; Cisło et al., 2004). The advantage of using carbon, as well as metals and metal oxides, is that their conductivities are independent of humidity (Zhang, 2011).

The conductivity of carbon fibres ranges between 10^5 and 10^6 S/m. Because they have a very high temperature resistance (up to 2000 ^{0}C in a non-oxidizing atmosphere), they may be used in situations where other materials such as copper and stainless steel are no longer applicable. In addition, their resistance against corrosive chemical agents and humidity might also be advantageous. However, their low mechanical shock resistance and abrasion resistance make them unattractive for use in smart textiles; their use remains limited. Carbon products: continuous fibres 100 % carbon (+grease), carbon+glass.

In general, carbon fibres are obtained by oxidation, carbonization and graphitization of organic precursor fibres. Polyacrylonitrile (PAN) is the most frequently used precursor fibre. The treatments produce a flexible fibre composed of 93-95 % carbon in a graphite structure (2D and/or 3D).

Carbon fibres appear as continuous multifilaments composed of 1,000 to 10,000 filaments with a diameter of 5 to 10 μm or as monofilaments covered with spin oil to improve processability and cohesion between the different monofilaments. Carbon-based yarns are processed into a textile structure through weaving or braiding.

Tanahashi et al. (1990), investigated the electrical resistances of materials produced by papermaking technology and used for manufacturing two electrically conductive line electrodes. In total, 65 % of this non-woven material was composed of resin-made carbon filaments, consisting of 20 % activated wood based on clay and 15 % polyethylene dispersing natural fibres used as a coupling agent. The electrical resistance of a non-standard product was 18 Ω.

Zhong et al. (1993) investigated two types of simply produced electrode materials commonly used in electrical elements. These materials were made in the form of graphite felt, made of polyacrylonitrile and artificial silk (viscose filament yarn). The authors argued that the electrical conductivity of graphite felt based on PAN is superior to that of artificial silk, despite the same structure of both felts.

In their work Cisło et al. (2004) aimed to reduce the specific vertical resistivity of active carbon non-woven materials, used in electrochemical conductors as an electrode material, in the initial production phase of viscose non-woven materials. The networks of active carbon non-woven materials have been produced in various ways. The vertical resistance of the material was measured using the (EN 1149-2: 2000) standard and the specific vertical resistivity (Ωm) was calculated by multiplying the value of the vertical resistance (Ω) by the ratio of the electrode surface area (m^2), used during the measurement, and the thickness of the tested non-woven material. The authors determined that the values of specific vertical resistivity of non-woven fabric depend on the method of manufacture of the fabric, for example, the specific vertical resistivity of the viscose non-woven fabric was 108 Ωm of queue; the specific vertical resistivity of the active carbon non-woven fabric, produced by the non-woven fabric technology, was 0.27 Ωm, and of the non-woven fabric, manufactured by classical seam technology, was 1.09 Ωm.

King et al. (2000) inserted conductive particles (black carbon and high-quality fine-grained graphite) into nylon at high temperature. The researchers found that black carbon improved the electrical conductivity of nylon (polyamide) because the chemical formula of black carbon is highly branched and carbon conductivity is better.

10.2.3. Conductive coated textiles

Besides the above methods, there are efforts targeting the modification of the common textile synthetic fibres in order to give them certain electrical characteristics. Some of the more important modifications are:

- Polymers filled with electroconductive compounds powder – usually carbon or metal powder is included in the mass of the polymer so that it permits the flow of electrical current through the fibres. Satisfactory results have been obtained when 25 % or more of the filling agent is added. Unfortunately, the presence of that amount of conductive particles strongly affects the mechanical properties of the fibres and limits the use of them in common textile applications. The fibres have a dark colour restricting their use in rather technical anti-static applications.
- Vacuum spread metal – the method targets the deposition of metal, e. g. aluminium, particles using a physical process. The particles are not strongly connected to the body of the fibre and the low adhesion gives poor results. The method is only applicable when a very thin layer of metal is required. If better electrical characteristics such as higher electrical conductivity are needed the metal coating cannot be easily obtained.
- Galvanic coating – the fibres are subjected to a galvanic process and are coated with a metal film. Although theoretically, it could give controllable results, the main drawback is that this method requires already electrically conductive fibres. Thus its application is limited mainly to carbon and graphite fibres.
- Chemical coating – this method is one of the most feasible methods for the production of electrically conductive textile fibres with good electrical characteristics. This chemical procedure is based on the treatment of the fibres in a bath, where metal salts are taken up by the fibres and then through the reduction, the fibres remain on the conductive metal. After the chemical coating, the fibres can be subjected to further galvanic metallization since they have the necessary electrical conductivity and homogenous distribution of the metal over them. The fibres keep the majority of their initial mechanical properties and they can be processed like the usual textile fibres without changes and modifications in the spinning process. The technological complexity of the method is the main disadvantage. The multistage process required, results in a relatively high cost of the conductive fibres. Another factor increasing the production cost is the processing of the wastewater

since after each stage the fibres must be carefully washed (Vassiliadis et al., 2004).

- Impregnation with anti-static agents – the cheapest, easiest and most commonly used method of providing the desired conductivity to textiles is impregnation of the material with anti-static agents (usually carbon filled resins). The result is an electrically conductive material, but the electrical properties are not stable and the conductivity not high enough. Another method of finishing is the direct coating of textile material with metals or other conductive substances. The most commonly used coatings are polypyrrole, polyaniline and polypyrene polymers – commonly named as inherently conductive polymers (ICP) (Dall'Acqua et al. 2004; Foitzik et al., 2007; Avloni et al., 2007; Marchini, 1991, Vassiliadis et al., 2004). This method gives satisfactory results but it introduces many drawbacks at the same time. Some of the positive aspects of this method are: (i) the electrical conductivity can be more precisely controlled; (ii) the production of the electrically conductive material is not complex; (iii) it is just one additional processing stage.

The long-lasting anti-static finish of textiles requires that the electrical properties of the surface be modified so as not to be seriously damaged by further washing or wearing. If the finishing reagent is applied to the surface, it must be firmly attached to the textiles and if the surface is chemically modified, that modification must remain intact throughout the life of the products (Nurmi et al., 2007; Hoback and Reilly, 1988; Dall'Acqua et al., 2004; Foitzik et al., 2007).

However, coating affects the structure of the fabric. The yarns and fibres are bonded together. During the use of the fabric, the deformations imposed affect the initial geometry of the structure. The inter-fibre and inter-yarn motion possibly break the continuity of the coating material resulting in the increase of the electrical resistance or in the discontinuation of the path of the electric current. Another problem is that this coating layer affects the colour of the textile material and consequently its appearance (Vassiliadis et al., 2004).

Hoback and Reilly (1988) examined the properties of graphite, silver, copper and nickel plating on the material surface and their surface durability. They found that carbon coatings were the best at controlling static electricity phenomena. Copper coatings were determined as well shielded, but they were not very useful for the environment due to the phenomena of oxidation. Graphite coatings had absolutely no shielding properties. The best shielding properties and electrical conductivity were found for nickel and silver. The resistances of nickel and silver were quite low, but these materials were the best at reflecting electromagnetic waves.

Xiao et al. (2014) presented three new methods for preparing electromagnetic (EM) textiles. These are metallized textiles containing pores and meshes, and spacer-structure textiles. They concluded that the SE of the metal mesh is obviously different from that of the metal plate with pores on its surface. For a plane wave in a distant EM field, the SE of the metal mesh depends mainly on the reflection loss. Most of the incident EM wave entering the fluff was reflected back to the receiver. Clearly, the EM waves were being absorbed by the plush fabric, based on the theory that the EM reflection coefficient is related to fibre size. The fibre size affects the incident impedance of the EM wave on the surface of the plush fabric and showed that the plush fabric exhibited frequency selectivity penetration. The velvet fabrics had less mass than absorbent materials at the same level of attenuation. In order to achieve the best scattering affect the conductive spacer yarns were grouped and each other arranged in a circular shape. On the lower surface of the spacer fabric, the spacer yarns were consolidated to give a smaller circumference, and on the upper surface, they formed a larger circumference, giving an upwardly open and relatively flat truncated cone. The EM wave could then be repeatedly scattered within the cone and the reflectivity in any particular direction would be decreased.

Nanocoating is a relatively new technique in the textile field. Conductive nanocomposite yarns are produced by various processes and techniques. Nanocomposite synthesis typically involves deposition of metallic nanoparticles or carbon nanotubes into a dielectric matrix. Polymeric matrices are of particular interest due to their relatively low cost and easy processability. The electric properties of such composites are closely related to the morphology of the embedded metallic or carbon nanostructures, which depends upon both film thickness and metal concentration (Raji et al., 2017).

Gupta et al. (2016) analyzed the influence of carbon (nano carbon black) concentration (5-10 %) in polyurethane resin coating on microwave properties of coated cotton fabric. The coating was applied using a knife-over-roll coating technique. EM range tested was 8-18 GHz. The coated materials were evaluated for relevant microwave properties: permittivity, scattering parameters, reflection, transmission, absorption, reflection loss and electromagnetic interference shielding in a VHS free space microwave measurement system. It was observed that increasing the carbon content increased the permittivity values and decreased the impedance. Tests results also revealed that with an increase of carbon content up to 7 %, absorption increased due to an increase of permittivity, while with a further increase of carbon content from 8 to 10 %, absorption decreased due to an increase of conductivity and thereby increased reflection.

Zou et al. (2015) analyzed the influence of washing procedures on the stability of shielding performance of cotton fabrics obtained via superhydrophobic finishing with Nafion (perfluorosulfonated polymer)/MWCNTs coating. According to the authors, the uniform distribution of MWCNTs is beneficial for not only forming a connective conductive network, which enhances electrical conductivity, and shielding performance of the cotton fabric but also constructing a nano-micrometre dual scale structure, which is necessary for the superhydrophobic surface. After 6-cycles of Nafion-MWCNTs deposition applying a drip-drying process, the resultant fabric possessed favourable shielding effectiveness of 9.0 dB. Besides the satisfied shielding ability and superhydrophobic surface, more importantly, the fabric exhibited good durability in EMI shielding after immersing in water for 96 h or washing in accordance with AATCC standard because of the superhydrophobicity and sufficient chemical stability of the Nafion-MWCNTs coating.

10.2.4. Inherently conductive polymers (ICP)

Intrinsically or inherently conductive polymers conduct electrical current which means that there are altering single (σ) and double (π) bonds between the carbon atoms of the backbone structure, and if it is doped (i. e. oxidation and reduction brings along electrons or supplementary holes moving along the chains) (Ramakrishnan, 2011).

The most frequently used ICPs are:

- Polyaniline (PANI)
- Poly-3,4-ethylenedioxythiophene (PEDOT)
- Polypyrrole (PPy)
- Polyacetylene

These polymers are insulating or semi-conductive in uploaded state but become conductive when doped.

Electrical conductivity to approximately 10 S/m may be achieved by adding dopants (<1 %). The amount of dopant added has a direct influence on the electrical properties of the polymer. The maximum conductivities that can be achieved reach up to 10^7 S/m, as in the case for metallic conductors.

ICPs are synthesized in a chemical or electrochemical way. They are infusible and difficult to dissolve, but they can be dispersed. These properties, together with the necessity to add dopants, make ICPs very delicate to handle. However, these materials are polymers, and are therefore lightweight and flexible (e. g. PANI has an elongation of up to 150 %).

In contrast with metals, ICPs are not resistant at high temperatures (e.g. 135 ^{0}C maximum for PANI).

Commercial products are available as a paste (i. e. particles dispersed in a binding agent). They can be coated (on fibres, yarns or fabrics) or printed (screen or inkjet) on the textile.

Two major disadvantages of these materials are:

- They are expensive
- They become unstable over time, due to their low resistance to humidity, oxygen and like all polymers, to temperature.

Yuping et al. (2006) found that the electrical conductivity of Pani-HCl composite was better than Pani-EB composite, as the ratio of Pani-HCl powder increases, the vertical resistance decreases. This was because when the quantity of Pani-HCl increases, "insulated conductive islands" dissipate and gradually merge with each other, i. e. a closed conductive circuit was formed in the composite.

The model multilayer shielding materials were developed (Brzeziński et al., 2009) for shielding electromagnetic radiation (EMR), showing a possible high absorption coefficient, a so-called insertion loss of ≥40 %. These materials were designed according to technology that consists of bonding/laminating constituent coating materials with diversified permittivity and magnetic permeability into multilayer systems. The individual layers were prepared using different technologies, including coating with ICPs–PANI, inserting metallized yarns, using conductive additives such as nano- and micro-carbon black, submicron and micro-powders of Al, Cu, Ni, submicron and micro-powders of ferromagnetic substances and semiconductors. Textile coating materials were prepared by the direct thin-layer techniques.

Polyaniline nanofibers and their composite with graphite have been synthesized by a simple chemical polymerization method (Joseph et al., 2017). Polyaniline nanofiber graphite composites with a thickness of 1 mm exhibit excellent electromagnetic interference (EMI) shielding of above 80 dB in the frequency range of 8.2–18 GHz. EMI shielding fabrics of 0.1 mm thickness based on polyaniline nanofibers as well as their composites have been developed by an in situ polymerization process. These fabrics combine the properties of polyaniline nanofibers, their composites and fabrics (cotton and nylon). The developed functional fabrics with 0.1 mm thickness exhibit EMI shielding effectiveness of approximately 11–15 dB in the 8.2–18 GHz frequency range. The SE_R value of the polyaniline nanofiber graphite composite is found to be in the range of 18–13 dB, while that of the polyaniline nanofibers is 14–10 dB in this frequency range. Similarly, with the addition of graphite, the SE_A value increases from 57–67 to 65–76 dB in the measured frequency range. It is found that the polyaniline nanofibers and their graphite composite exhibit an attractive EMI shielding response for a small sample thickness of 1 mm with a significant contribution from SE_A.

Novel conductive fabrics were developed by polymerizing aniline onto the polyamide (PA)-knitted fabrics (Engin and Usta, 2015) The highest shielding value obtained was 6.7 dB in the frequency range from 2200 to 2300 MHz in the 0.5M aniline-concentration treated fabric. The absorption, reflection and shielding values were found to be better for 0.5M aniline concentration as compared to the other treated fabrics. It is also observed that the increase in the absorption values directly affects the shielding values and the fabrics perform the absorption instead of reflection.

Järvinen and Puolakka (2003) found that the conductivity of filter polyester fabrics can be improved by the influence of the filter material on polyaniline. The conductivity of the affected polyester fabric remains unchanged, even in extreme weather conditions during transportation, storage or cleaning. The authors exposed polyaniline-treated polyester fabrics to moisture, cold and UV (ultraviolet) light in order to measure possible changes in the electrical conductivity of the filter fabric. The researchers found that at -20 ^{0}C, the conductivity of the fabric remained unchanged, and when the fabric was treated with drops of water or fully soaked, the electrical conductivity was not altered very significantly. The test fabrics tolerated impregnation in distilled water for seven days well, i.e. the electrical conductivity did not change significantly either. Even after 100 days of continuous 60 % humidity exposure, the vertical resistance (measured by the EN 1149-2 method) was still approximately 10^5 Ω lower than the untreated with polyaniline polyester fabric. The intense UV light has the greatest effect on polyaniline-treated fabrics, although the loss of electrical conductivity remains for two decades.

10.2.5. Textiles with conductive fillers

Polymers are mainly insulating materials. However, by incorporating conductive fillers into an insulating polymer matrix, electrical conductivity can be achieved. The type and amount of conductive filler like the combination of the filler and polymer matrix depend on the required electrical and mechanical properties.

Three main types of conductive fillers are applied:

- metallic particles (copper, silver, nickel, stainless steel) are used for their excellent electrical conductive properties;
- carbon particles (graphite, carbon black, carbon fibres and carbon nanotubes) are used for their mechanical reinforcing properties;
- intrinsically conductive polymers are used for their low density and their mechanical properties that resemble those of the hosting polymer matrix.

Polymer matrices are mainly categorized into three groups:

- Thermoplastics: easy to shape and reusable;
- Thermoset: very stable but difficult to reuse;
- Elastomers: stable, cross-linked, highly elastic but not recyclable.

To produce products with conductive fillers, several "mixing" methods are applied:

- In situ polymerization and vulcanization is used for thermoplastics and elastomers. The charge may be introduced into the polymer matrix before or during the polymerization or cross-linking process. But always before the final shaping of the material.
- The melting process (mixture or extrusion) is used for thermoplastics. It is carried out to perform a pre-mixture, followed

by shaping the material through compression, injection or extrusion).
- The wet process is used to increase the dispersion of the charges and, in some particular cases, the implementation (e. g. unmeltable polymers).

The nature and morphology of the filler influence the percolation concentration – the minimum amount of filler that needs to be added to create a conductive polymer composite. Thus, it is important to find a charging material that can introduce conductivity into the polymer by just adding a low concentration.

In theory, the ideal percolation concentration for a perfect dispersion of spherical conductive particles is 16 % in volume.

In reality, however, concentrations are:

- 14 to 20 % for graphite (spherical particles or platelets);
- 7.5 to 15 % for carbon black (agglomerated or ramified spherical particles);
- 0.5 to 3.5 % for carbon nanotubes.

Because of these important disadvantages, conductive polymer composite cannot be produced in fibre form (except for those filled with carbon nanotubes). Therefore, they are used for coating and printing.

Rubežienė et al. (2015, 2018) investigated conductive fabrics, where metallized yarns were interwoven in 85 % cotton/15 % polyester fabrics and formed a mesh grid (grid area 50 mm^2 and 100 mm^2) and where conductive coating contained conductive filler carbon black as an absorber of EM waves was applied. S-Shield PES® yarns produced by Schoeller GmbH&CoKG (Austria) – Polyester (PES) fibres with stainless steel staples (INOX) – 80 % PES, 20 % INOX were used. The diameter of the metal fibres was 8 μm. In the case of coated fabrics, the carbon black/styrene acrylate copolymer coating was applied manually, using a screen printing method. The amount of carbon in the paste was determined at 24.12 %. The samples of fabrics were printed with the particular paste coating their full surface (50x50 cm) or using the squared printing pattern, with square dimensions of 1 x 1 cm and lines of 2 mm width. In order to bond and fix the conductive layer on the fabric, the samples were dried in the laboratory oven and steamer. EMR shielding effectiveness in the frequency range 2–12 GHz of investigated samples was evaluated using a procedure based on measurements of reflection and transmission of electromagnetic waves, normally incident on fabric, in a semi-anechoic chamber. The EMR attenuation of the developed samples over the frequency range 2–20 GHz was evaluated using a UWB Time Domain measurement system. It was determined that shielding effectiveness decreases with frequency for samples with silver-coated fibres. For investigated fabrics with metallized yarns, the contribution to shielding from absorption and reflectance is similar in the case of a sample with finer mesh. For another sample – fabric with 100 mm^2 mesh grid, the reflectance is weightier. Moreover, the shielding properties of fabrics with metallized yarns in all tested ranges are changeable. It was determined that the distribution of coating paste on the fabric's surface had an important influence on their EMR shielding effectiveness at the higher frequency range. Fabrics with partial coating exposed considerably lower shielding effectiveness (SE) within a frequency range of 2-20 GHz, in comparison with fully coated fabrics, due to the open space (not coated) structure, allowing EMR to pass through the structure at shorter wavelengths. It was also determined that for carbon-coated fabrics absorption contribution to shielding was the deciding factor.

11. CONDUCTIVE "SMART" AND E-TEXTILES

The field of conductive textiles may be explained as an integration of technologies of materials, electronics and textiles in order to create a new generation of flexible, comfortable, multifunctional conductive textile structures. This structure is known as 'Smart textile' (Anderson and Seyam, 2004). Smart Textiles covers a broad field of studies and products that extend the functionality and usefulness of common textile materials. Smart Textiles are defined as textile products such as fibres/filaments, yarns, woven and knitted fabrics, non-woven structures, which can interact with the environment or user. The interflow of textiles and electronics (e-textiles) can be important in developing smart materials that are capable of accomplishing various functions found in rigid and non-flexible electronic products nowadays.

Smart textiles integrate a high level of intelligence and can be divided into three subgroups (Van Langenhove and Hertleer, 2004; Van Langenhove et al., 2005; Chudasama, 2014):

- Passive smart textiles: only able to sense the environment or user and are based on sensors;
- Active smart textiles: reactive sensitivity to the surroundings, integration of an actuator function and a sensing device;
- Very smart textiles: able to sense, react and adapt their behaviour to the given circumstances.

The percentage of e-textile players using each material type, derived from IDTechEx's survey of over 150 suppliers and manufacturers in the space is presented in Figure 11.1.

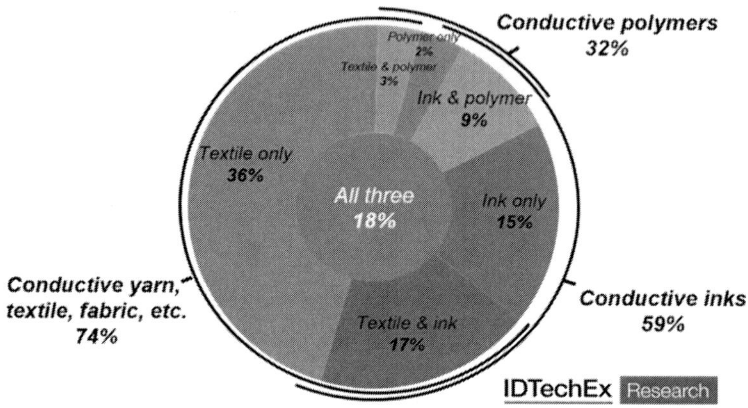

Figure 11.1. Number of e-textile players that use various conductive materials in 2018

Some of the main players dominating in the e-textile industry are E. I. Du Pont De Nemours and Company, Intelligent Clothing Ltd., Interactive Wear AG, International Fashion Machines Inc., Kimberly-Clark Health Care, Milliken & Company, Noble Biomaterials Inc., Outlast Technologies Inc, QinetiQ North America, Royal Philips Electronics N.V., Toray Industries Inc., and others. This industry is classified based on the following application segments: Consumer Products, Military & Homeland Defense/Public Safety Applications, Computing, Biomedical, Vehicle Safety & Comfort, Others (Logistics & Supply Chain Management, and Signage, among others). Major geographic areas include North America, Asia Pacific, Europe, and Rest of the World (www.transparencymarket research.com).

Different fabric sensors, fabrics incorporating thermocouples, luminescent elements, wearable displays, shape-sensitive fabrics, carbon electrodes and other materials may be attributed to Smart or E-textiles. In all these products the presence of conductive elements in textiles is critical. While manufacturing E-textiles, the conductive materials are incorporated into the textile structure by different technologies, e.g. embroidering, sewing, non-woven textile, knitting, weaving, spinning, braiding, coating/laminating, printing and chemical treatments (Stoppa and Chiolerio, 2014). But, Smart or E-textiles cannot function properly if the electric circuit is not formed properly in the fabric by conductive materials.

There are primarily two categories of E-Textiles (Wainwright, 2016):

1. The first and simplest applications are embedded "off-the-shelf" consumer electronic devices, such as digital cameras, battery packs, chargers, speakers, headphones, mobile phones, and light emitting diodes (discrete LEDs and LED strips, flexible solar films, digital audio players, EL wires, etc.). These are installed into apparel using standard wiring.

2. The second and most compelling category includes E-Textiles that:

- Employ micro-processor-controlled fabric colour with changing systems that are remotely controlled by sensors or smartphones;
- Monitor and display real-time biophysical conditions in the form of digital alphanumeric data; - Replicate physical motion or mood remotely on another person's apparel;
- Provide responsive colour-changing attributes to fabric, depending on a person's mood or in response to surrounding stimuli (i.e., sound, temperature, light, lethal gas environments, etc.);
- Use fabric displays to identify incoming callers on nearby smartphones.

Advantages of E-Textiles are somewhat obvious when considering utility value-added technologies that can be combined with textiles (Wainwright, 2016):

- monitoring health and fitness (the largest segment of R&D),
- recharging digital devices fitted with lightweight batteries such as LiPo (Lithium Ion Polymer batteries), by using body heat or body movement,
- providing night-time LED safety indicators,
- IFF (Identifying friend or foe) in policing or combat situations,
- indicating acute surrounding environmental hazardous conditions (i. e. UV radiation levels),
- displays that link to smart-phone functions, such as emails, texts, voice mails, calls, GPS,
- alerting hospital and first aid staff to medical emergencies,
- using smart-phone-controlled fabric attributes to display utility functions such as voltage, temperature, and other measurements.

11.1. The scheme of the electric circuit in the textiles

Electrically conductive yarns and fabrics are widely used where properties such as flexibility and comfort are important. Conductive fibres have been used for quite some time in the manufacture of electromagnetic shielding and electrostatic dissipation fabrics. Also, conductive materials are used for the manufacture of e-textiles or as electrode and battery separators (Kim et al., 2004; Dhawan et al., 2004).

The simplest way to embed conductive yarn into the fabric is to weave it as one of the warp or weft yarns (El-Newashy et al., 2012). Furthermore, one can go for modern shuttleless weaving technology. Warp and weft knitted structures are also possible and can give different effects than those of woven fabrics as far as physical characteristics are concerned. Sometimes it is enough to form a conductive yarn net structure or insert conductive yarn only in one direction in the fabric structure in order to receive good EMI shielding properties or protection against incendiary discharge.

One of the most noteworthy discoveries in the use of conductive yarns in electronics is the attempt to form a yarn mesh in the fabric during weaving or knitting (see Figure 11.1.1). Such fabric can be described as a network of conductive and non-conductive yarns arranged and weaved in accordance with the prior fabric design. Conductive fibres behave like an electric signal carrier in this network that transmits signals from one point of the network to another. An efficient current transfer from a conductive yarn to the next perpendicularly weaved-in conductive yarn in a weaved electric circuit requires an efficient electrical connection (the point of interconnection). The main goal of the interconnection point is to reduce the vertical and surface resistances associated with the interconnection point that is achieved by selecting the right material and the particular connecting/weaving procedure (Paasi et al., 2004; Paasi et al., 2005; Dhawan et al., 2004; Ghosh and Dhawan, 2006; Ouyang and Chappell, 2005).

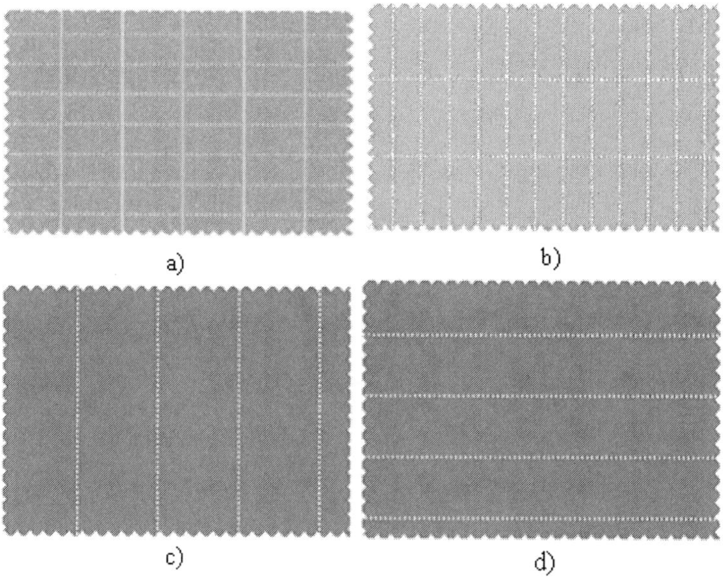

Figure 11.1.1. Techniques for inserting conductive yarns into fabrics: a, b – mesh structure; c, d – is a one-way structure for the insertion of conductive yarns.

Pinar and Michalak (2006) investigated knitted fabrics of different weaves, in which a VSC Lenzing 75%/PES 25% two-component electrically conductive yarn was inserted in a transverse direction at a distance of 10 mm. The polyester component of the yarns was composed of Resist type carbon particles. The authors conducted the experiments according to European standards (EN 1149-1: 2000; EN 1149-2: 2000; EN 1149-3: 2004). They found that the surface resistance of knitted fabrics with the highest yarn filling coefficient is the lowest, and the charge decay time is the shortest, i.e. this fabric has the best electrostatic properties. They also found that values of shielding factor only depend on the number of conductive yarns knitted in the fabric – the more conductive the yarns, the better the shielding properties of the fabric. The resistance values of the investigated materials were at the 10^{10}–10^{11} Ω degrees.

Su and Chern (2004) explored the effectiveness of shielding fabrics where stainless steel yarns were woven in at various distances in the directions of weft or weft and warp. They found that fabrics in which conductive yarns were woven into fabrics in both directions and formed a fine mesh, shielded much better than fabrics in which conductive yarns formed a larger grid. The authors also established that the worst shielding properties were found in those fabrics where the conductive yarn was woven in only the weft direction. Studies have also shown that the mesh size, formed by conductive yarn in plain weave fabrics and 2/2 twill weave is small and the mesh size of fabric with the 3/1 twill weave is narrow (the area of conductive mesh was found to be larger for this fabric than for fabrics with the plain weave or 2/2 twill weave). Therefore, the electrical conductivity of the latter fabric is lower compared to the other two materials. The shielding efficiency of plain weave fabrics was found to be the best. The statement that the conductivity of the fabric depends on the mesh size formed by the conductive yarns is questionable because the main factors affecting the properties of the fabric are the yarn densities and their lengths.

Lin and Lou (2003) found that the molten four-layer non-woven polypropylene strip forms a thermoplastic laminate matrix. The conductivity of laminates after the thermal process equals the conductivity of the untreated laminates. Although conductive metal multifilament yarns were inserted into the laminates only in the direction of the weft, and after heat treatment the polypropylene non-woven material shrank and became more compact, the surface resistance in all directions of the laminate became uniform.

A cut of electrically conductive textiles – woven or knitted, can be treated as a combination of resistors that correspond to the resistance of the yarns themselves and the resistance at the points of contact among yarns on both sides of the fabric (contact resistance). These parameters depend on many external factors such as temperature, humidity, pressure and elongation (Banaszczyk et al., 2009).

The conductivity of electrically conductive material can be assessed by an electrical scheme. The loop model shown in Figure 11.1.2. consists of resistance (R_c) at the contact points of yarns, a loop float resistance (R_l), and tuck loop resistance (r_l), and related respectively with an overlapping load force of the yarn and an internal resistance of the metal yarn (Zhang et. al., 2005; Tao, 2005).

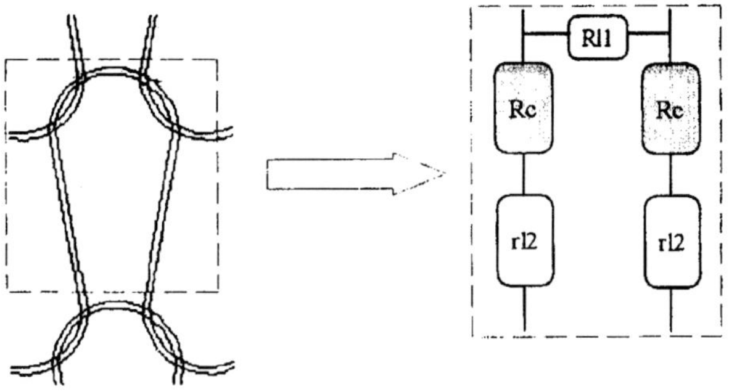

Figure 11.1.2. The loop and its simulation model using an electric resistance circuit (Banaszczyk et. al., 2009).

Loops change their shape when fabrics are under load. However, the contact resistance R_c and resistance R_l, which is dependent on the length, will vary depending on the size of the load force and the length of the yarn in the loop that will change when the fabric is stretched.

The equivalent resistance of the fabric is calculated as the ratio of voltage U to total current i_{total}.

$$R_{equivalent} = \frac{U}{i_{total}} \tag{39}$$

Kirchoff's circuit law and electrical circuit theory are used to calculate the fabric circuit i_{total}. Kirchoff's circuit law shows that the sum of all component voltages in a closed circuit is equal to zero. The hypothetical current i flows through each closed circuit circulation, and the true current I flowing through the circuit branches is the arithmetic sum of the hypothetical current i (Zhang et. al., 2005).

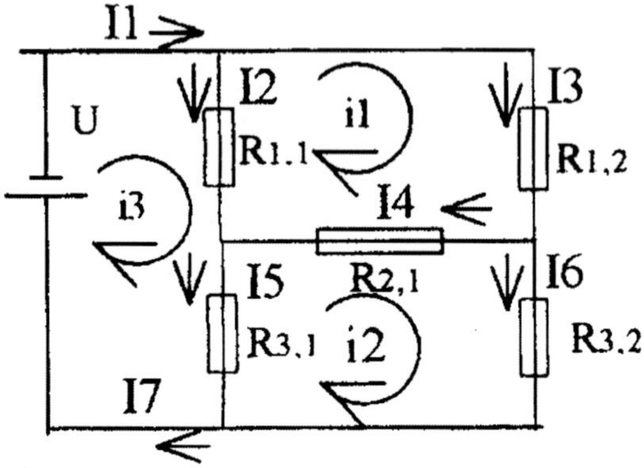

Figure 11.1.3. Loop circuit model (Zhang et. al., 2005).

The loop circuit model in Figure 11.1.3 is analyzed according to the method described above. The relationship between the hypothetical current *i* and the branch current *I* can be expressed by the formula (40). The formula (41) is derived from Kirchoff's circuit law.

$$\begin{cases} I_1 = i_3 \\ I_2 = i_3 - i_1 \\ I_3 = I_1 \\ I_4 = i_1 - i_2 \\ I_5 = i_3 - i_2 \\ I_6 = i_2 \\ I_7 = i_3 \end{cases} ; \qquad (40)$$

$$\begin{cases} I_3 R_{12} + I_4 R_{21} - I_2 R_{11} = 0 \\ -I_4 R_{21} + I_6 R_{32} - R_{31} I_5 = 0 \\ I_2 R_{11} + I_5 R_{31} = U \end{cases} ; \qquad (41)$$

After solving these equations, we would find that the equivalent resistance of the loop circuit model presented in Figure 11.1.3 could be calculated from formula (39).

The equivalent resistance of any conductive knitted material can be calculated using the provided solution model (Zhang et. al., 2005).

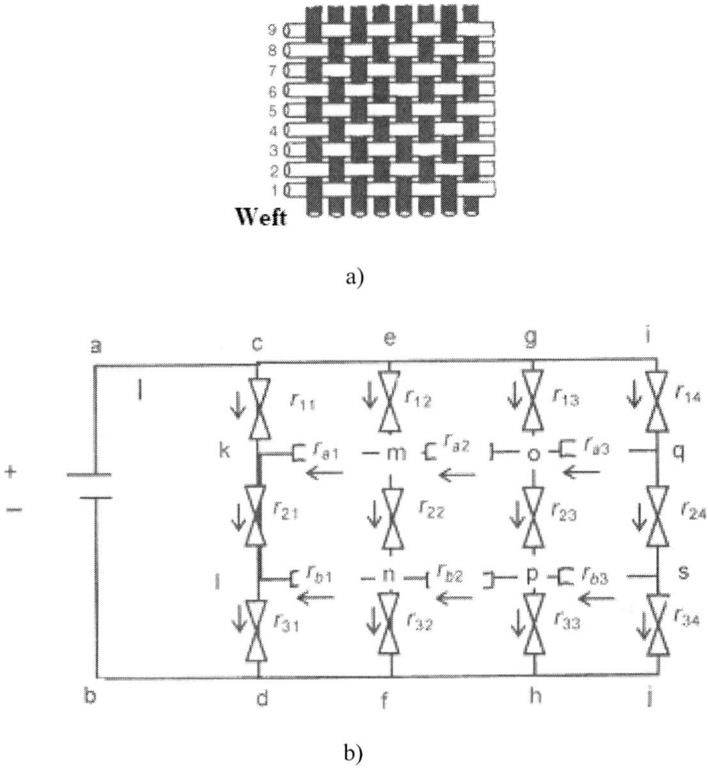

Figure 11.1.4. a) Structure of plain weave woven fabric; b) an electrical circuit describing a segment of conductive woven fabric (Zhang et. al., 2005).

The plain weave fabric coated with a conductive polymer can be described by the electrical circuit shown in Figure 11.1.4. The resistance of the plain weave fabric can be expressed (Tao, 2005):

In warp direction: $$R_v = \frac{\lambda(1+C_v)(N_p-1)}{N_e};\qquad (42)$$

In weft direction: $$R_h = \frac{\lambda(1+C_h)(N_e-1)}{N_p};\qquad(43)$$

where: R_v, R_h– the resistance, measured in warp and weft directions; N_p– weft density; N_e– warp density; λ– yarn resistance through total length; C_v and C_h– weave overlay in warp and weft directions, respectively (Tao, 2005).

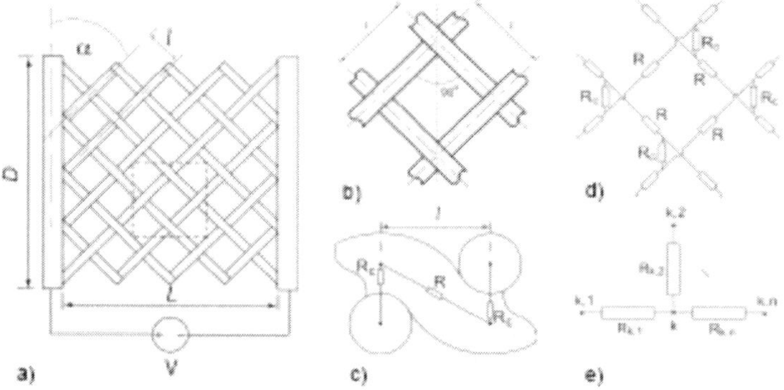

Figure 11.1.5. Electrically conductive textiles; a) overall view, b) magnified fragment of woven fabric, c) scheme of contact resistance, d) an equivalent electrical model is woven for the fragment of woven textile, e) explanation of recurring method (Tao, 2005)

The scheme of woven electrically conductive textiles is shown in Figure 11.1.5. The angle α between the fibre and electrodes is called the angle of contact. The resistance of the cut R_{cut}, Ω, is calculated (Banaszczyk et. al., 2009):

$$R_{cut} = \frac{V_{el}}{I_{el}}\qquad(44)$$

where: V_{el} and I_{el}– voltage and current between electrodes. The value of R_{cut} depends on the shape and dimensions of a specimen. The specific surface resistivity of cut ρ_{cut} is calculated:

$$\rho_{cut} = \frac{\frac{V_{el}}{L}}{\frac{I_{el}}{D}}\qquad(45)$$

where: *L* and *D*– the length and width of the cut, respectively (Figure 11.1.5.a.), ρ_{cut}– specific resistance, Ω, which describes the fabric regardless the size and shape of the specimen (Banaszczyk et al., 2009).

Electrically conductive textiles can be interpreted as an electrical circuit consisting of resistors connected in parallel and sequentially. A magnified fragment of conductive textiles is shown in Figure 11.1.15.a. and the corresponding electrical model is shown in Figures 11.1.15.d and e. The electrical resistance of the part of the fibre whose length *l* (the distance between the two parallel yarns) is indicated as *R*. The contact resistance between interlaced yarns is called R_c. Both of these resistances are determined experimentally for each of the fibres tested. The resistance values vary considerably depending on the material from which the yarn was made, the structure of yarn (e. g. they consist of many thin fibres), as well as external factors such as the load between the contacting fibres or the moisture of the textiles (e. g. the impact of human sweat). The idea of an electric model is simple, but the main problem is a large number of circuits that can reach up to one hundred thousand elements, which makes it difficult to calculate the results using a simulation product electrical scheme. An analytical solution using Kirchoff's voltage and the current law is not possible due to the large number of unknowns. Banaszczyk et al. (2009) developed a computer program where the new recurring potential of each intersection point is calculated using the formula:

$$\emptyset_k^{new} = \emptyset_k^{old} + \tau\left(\frac{\frac{\emptyset_{k,1}}{R_{k,1}}+\frac{\emptyset_{k,2}}{R_{k,2}}+\cdots+\frac{\emptyset_{k,n}}{R_{k,n}}}{\frac{1}{R_{k,1}}+\frac{1}{R_{k,2}}+\cdots+\frac{1}{R_{k,n}}} - \emptyset_k^{old}\right) \quad (46)$$

where: $\emptyset_k^{new}$– a new potential at *k* intersection point after repetition, $\emptyset_k^{old}$– a potential at *k* intersection point before repetition, τ - relaxation coefficient, $\emptyset_{k,1}$, $\emptyset_{k,2}$ ... $\emptyset_{k,n}$– k a potential of intersection points of neighbouring intersections, R_1, R_2, ...R_k resistances of branches connected at the intersection point *k* as shown in Figure 11.1.5e, which may be equal to *R* or *Rc*.

In his work (Dhawan et al., 2004), A. Dhawan and co-authors investigated the contact resistance of the copper yarns. They found that the resistance *R*, measured in multimeters, actually consists of the sum of many components. The fibres are not continuous and within the yarn structure are made of many fibres, the superposition principle is not valid. The electrical continuity is not due to the mechanical continuity of the fibres-conductors. It is achieved through a combination of alternating fibres being in contact. The total resistance is the sum of the resistance of the fibres plus the contact resistance between the fibres (see Figure 11.1.6) The resistance of the fibres R_f - a character of the material. The metallization process is controlled and can provide certain specific resistance depending on the metal content of the fibres. The contact resistance R_c depends mainly on the structure of the yarn and on the mechanical conditions within the structure. It is a function of the area of contact and of the applied pressure between the yarns. In the conductive path the total resistance is (Vassiliadis et al., 2004):

$$\mathrm{R_{tot}} = \sum_{i=1}^{n} \mathrm{R_{fi}} + \sum_{j=1}^{m} \mathrm{R_{cj}} \tag{47}$$

Meanwhile, while measuring the resistance of the non-weaved conductive yarn, these components are not affected. The resistance measured at the point where the yarns are connected includes other resistances (which may be called *Ro*), such as the resistance of the multimeter probe, the resistance of the multimeter probe connection to the conductive yarn (crocodile terminals), and so on. In characterizing the connection efficiency of conductive yarns, Ro should be subtracted from the measured *R* value.

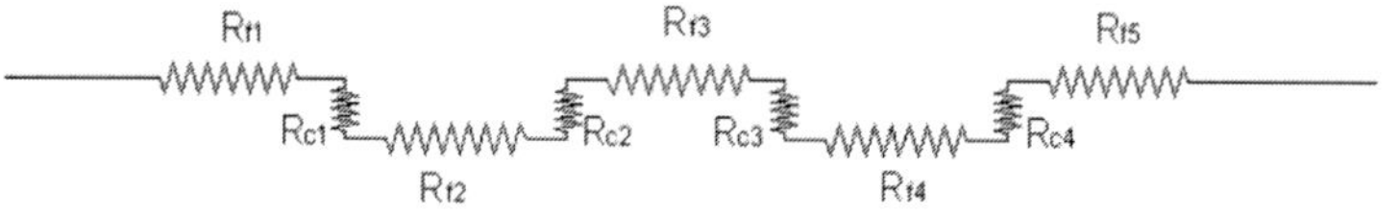

Figure 11.1.6. Arrangement of fibres and its electrical equivalent (Vassiliadis et. al., 2004).

11.2. Wearable intelligent textile systems

The general principles, definitions and categories of smart textiles are described in document CEN/TR 16298. The main components of the wearable intelligent textile system (ITS) are an actuator associated with a specific purpose sensor, an electronic information controller (processor), an energy source, and a communication device. An example of the simplified principle scheme of ITS is presented in Figure 11.2.1.

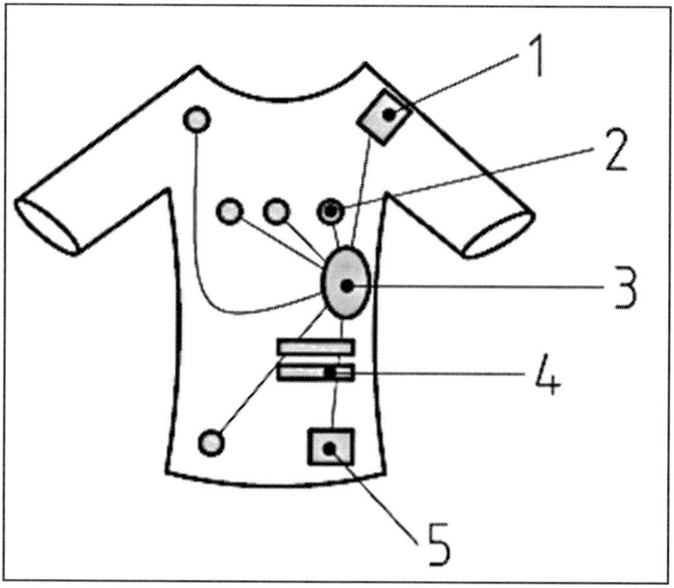

Figure 11.2.1 An example of a wearable intelligent textile system: 1 – communication; 2 – sensors; 3 – processors; 4 – actuator; 5 – energy supply (CEN/TR 16298)

Modern intelligent clothing, as compared to the traditional one, has additional active features. These features provide the use of textile materials with unique properties, or integrated wearable electronic devices, or a combination of both (Stoppa and Chiolerio, 2014; De Mey et al., 2014). According to the nature of intelligence, clothing can be divided into three levels (Srazdienė and Dobilaitė, 2007):

- clothing -"assistant" who collects and stores certain information and processes the data received;
- clothing -"observer", which records human physiological parameters, follows the state of health;
- clothing-"regulator", which tracks and adjusts the required parameters. This level includes heating textile products.

Currently, the heating of smart clothing products based on non-textile electronic components (called "wearable electronics") already exists on the market, but these are not yet widely available to consumers.

It is known, that a prototype of a heated, intelligent clothing product was developed in Finland at Tampere University of Technology in 2006 (Vargas, 2009). The base of the product is a heated shirt with digital temperature sensors that measure the wearer’s skin and a glowing carbon fibre panel, powered by high-capacity nickel batteries. The heating temperature range can be chosen by the wearer himself. The electronic circuitry and the heating panel must be removed before washing the clothing. The heating element made of carbon fibre is also used by the Milwaukee “M12 ™ Heated Gear” in the production line of protective clothing against foul weather. These products have 3 heating capacities; Li-ion batteries have a battery life of 3.25 hours (the lowest heat-up time) up to 8 hours (the highest heat-up time).

The North Face company's heating jacket MET5 is also known in the market. It has 2 heating zones formed from conductive fibres and maintains the optimum temperature for the wearer by pressing the textile button. This technical solution made by both Malden Mills and North Face produces a heated jacket called Polartec® HeatTM, which uses a heat-generating network of thin fibres fused to the chest area. The entire heating system is powered by two rechargeable Li-on batteries. There are 2 heating modes available – maximum (up to 45 °C, duration 2.5 hours) and variable (duration – 5 hours). All electronic components must be removed before washing.

Heated underwear is manufactured by WarmX GmbH. The main idea of Warm-X technology is to achieve the warmth effect with the lowest possible energy consumption. According to this concept, heating elements are incorporated into underwear, where direct skin contact minimizes energy loss. These garments are produced from knitted fabrics made of conductive silver-plated polyamide yarns. Constant heating power (optionally 2, 4.5 or 7 W) is achieved by the microprocessor and lithium-ion batteries (continuous operation: power level 1– up to 6 hours, 2– for about 4.5 hours, 3– for about 2.5 hours).

Textile heating garments (underwear, gloves, socks, waistcoats) are also advertised by such companies as Duran, China Depot, Zanier-Sport GmbH, Inuheat, and others.

Researchers at Cornell University (USA) assessed that the maximum temperature attained by the heating elements that come into contact with human skin must not exceed the tolerance limits of the human temperature – 40 °C (Vargas, 2009).

Šahta et al. (2014), concluded that steel yarns could not be used for knitting while silver-coated polyamide yarns are suitable for the manufacture of knitted heating elements. Depending on the conductive yarn type inserted in the fabric, it can increase or decrease fabric resistance in relation to initial resistance. Evenness of warm-up was provided by three-ply silver-coated polyamide yarns. The knitted heating element of three-ply yarn performed the heating functions without any significant changes in temperature over a long period of time. The main criteria for the selection of yarns for manufacture of the knitted resistive heating element are low electrical resistance of yarns; low resistance changes during the elongation of the knitted heating element; constant temperature of the element for a protracted time and the suitability of yarns for knitting.

A comparison of heat generation by plain, rib and interlock structures was studied in the scientific article by Hamdani et al. (2013). It was discovered from a number of various experiments that a minimum required threshold force of contact at binding points in knitted fabric exist for the electricity to pass. Once this force is achieved, stretching the fabric does not affect the amount of heat produced.

Bai et al. (2018) presented a unique and facile technique for fabricating flexible heating materials with temperature perception; a temperature-sensitive fine copper was integrated into two pieces of flexible fusible interlining fabrics by a simple thermal bonding method. The electrical properties, thermal performance and mechanical properties of investigated materials are stable according to the experimental data and analysis presented, and it may be helpful for developing secure and durable products.

Investigations were carried out to create a textile heating circuit and to adopt the required technical parameters for "wearing electronic" components (sensors, centralized sensor/heating device connectors, temperature controller, power supply and flexible conductive connections) while designing and developing intelligent heating clothing with wearable electronic, that actively react to human body/environment temperature changes and intended to use in active physical activity in a cold environment. The operation of designed wearable ITS requires a coherent interaction between all the above components.

The first "wearing" electronic component needed for warm clothing – a sensor, measuring and following the wearer's skin temperature and transmitting this data to the controller. In choosing temperature sensors, it is first and foremost important to define the main measurement indicators: sensitivity and accuracy. Integrating the temperature sensor into a textile material, it is also important for it to be small and securely fastened; its transmission highway must be flexible, have a minimum number of threads, be as short as possible and isolated from external influences.

The temperature of the human skin, depending on the environmental conditions, varies in different parts of the body– 32–37 °C (Pan and Gibson, 2006). The average human skin temperature is assumed to be about 34 °C (Koralewski, 2006). Exposure to the skin above 40 °C for a warmer surface for more than 30 seconds has a potentially painful effect. Table 11.2.2 presents the conditions for the thermal sensation and human thermophysiological comfort depending on the temperature of the skin.

Table 11.2.2. Dependence of the thermal sensation on the wearer's skin temperature

Thermal sensation	Temperature of human skin, °C
Very hot	>36.6
Hot	36.0±0.6
Warm	34.9±0.7
Comfortable	33.2±1.0
Cool	31.1±1.0
Cold	29.1±1.0
Very cold	<28.1

Temperature control is one of the most important functions of clothes. Most of the heating elements use the principle of Joule's heat, which is generated when an electric current is passed through a conductive material. All conductive materials are heating elements in principle. The basis of heating textile products consists of materials that have electrical conductivity properties. Analyzing technical literature and the market for this type of product (CEN/TR 16298; Stoppa and Chiolerio, 2014; De Mey et al., 2014; Strazdienė and Dobilaitė, 2007; Vargas, 2009; Šahta et al.,

2014; Hamdani et al., 2013; Locher, 2006; Sezgin et al., 2012; Ding et al., 2014; Roell, 1996; Lee Sandbach et al., 2008; Petcu et al., 2012; Poboroniuc et al., 2014; Mečnika et al., 2014), it can be concluded that in recent years clothing products have been designed to integrate heating elements in the material structure, which are composed of conductive yarns introduced in various ways.

Electroconductive textiles cannot be considered as homogenous structures, because fabrics consist of conductive and non-conductive yarns, interlaced with each other. Such distribution of yarns results in anisotropic current distribution when a voltage is applied. The aim of investigations made by Varnaitė-Žuravliova et al. (2016) was to assess current and temperature distributions in conductive textiles, which can be used in many applications, such as protective textiles, e-textiles, heating textiles etc. It was found that Ohm's law is valid for such type of textiles and temperature increases the increasing voltage applied. The quantity and value of current passing through the conductive yarn principally depends on the conductivity of the yarn. The length of the conductive yarn also influences the values of current and temperature. It was found that because of the continuous coating the electric current passes through yarns with silver-coated filaments more homogeneously than in yarns with metal fibres. It was also found, that there exists a maximum voltage which can be fed to the conductive yarns, without damaging them.

Petcu et al. (2012) aimed to make a comparison between five different types of conductive, heatable samples. These textile samples were produced according to the five most important implementation techniques: knitting, weaving, embroidery, inkjet printing and non-woven padding. The idea is to identify a conductive option best suitable for a heating application. The first three methods use electro conductive wires as heating elements, the fourth uses conductive ink and the fifth uses carbon black coating. For all of them, resistance, current and heat distribution were measured. The inkjet printing technology showed an excellent heating behaviour (up to 105 °C with an applied voltage of 14 V) on a polymeric substrate, but when it came to a textile substrate, the results were strongly influenced by the structure of the substrate and its shift during stretching. Conductivity was achieved even after repeated stretching but in special processing conditions. Using non-woven carbon padded fabrics as heating elements proved very successful as well. Its lightweight and reduced thickness play an important role in the processability of the actual heating system. The downside of this technology is the reduced elasticity of the fabric. In addition to the reduced elasticity, the fabric is very resistant and it requires a high voltage to reach the desired temperature range (40 V for 80 ˚C). The last three implementation methods, knitting, weaving and embroidery, were based on the same type of heating element – electroconductive textile threads. Embroidery proved to be the most reliable technology, even on the elastic substrate, reaching a 78 ˚C temperature with a voltage of 25 V. The samples performed better thermally but unfortunately, not all conductive textile threads are embroidered.

Sezgin et al. (2012) examined conductive yarns used as transmission lines of e-textile structures due to the increase in voltage. In total ten different plain fabric samples were produced: five different conductive yarns with different linear resistance values were used in order to form an e-textile structure. As for the conductive yarn, three different silver yarns and two different steel yarns with different linear resistance values were used. Then, the voltage was applied through the textile circuit from the end of the conductive yarns and their thermal images were captured by the thermal camera. Results showed that the temperature along the conductive yarn due to an increase in voltage value varies according to the type of conductive yarns and the base yarn of the fabric. The maximum voltage was obtained with the 70-denier silver yarn which has a linear resistance of 377 Ω/m. The minimum voltage was obtained with 526 tex steel yarn which has a linear resistance of 17 Ω/m. Additionally, due to the thermal conductivity coefficients of fibres, it was found that 100 % of cotton samples reach higher temperatures than the 50 % cotton/50 % acrylic samples. So, it is obvious that instead of fabrics composed of 100 % cotton yarns, using fabrics composed of 50 % cotton/50 % acrylic yarns is more convenient for e-textile applications. Finally, using 70-denier silver-plated nylon yarns with 50 % cotton-50 % acrylic yarns is more suitable for e-textile applications since they present better (lower) temperature values due to an increase in voltage value.

The electrical resistance of knitted fabrics embedded with conducting yarns at different temperatures was studied by Ding et al. (2014) in their paper. Two types of resistance: linear and contact resistance have been investigated. Two kinds of silver-coated conductive yarns, with linear resistance of 68.6/cm and 1/cm, were embedded into normal knitted woollen fabrics. The temperature impact on the resistance of these two conductive woollen fabrics as a function of applied voltages was investigated. The results showed that the resistance of every conductive fabric decreases by a maximum of 30 % when the temperature is rising. This can be explained by two basic factors: the electrical resistance of the silver-coated conductive yarns decreases as the temperature rises; the physical contact of the overlapped conductive yarns elongates with regards to heating on knitted fabrics, which causes a decrease in contact resistance. Not only the linear resistance of the conducting yarns but also the contact resistance between conducting yarns, play an essential role in the total electrical resistance of the conductive knitted fabric. They both decrease with increasing temperature. In summary, the total resistance of the conductive knitted fabric changes dramatically when the fabric is heated.

There are two ways to manufacture heated textiles (Hamdani et al., 2013):

1) produce a technical fabric and then integrate electronic components;
2) produce a technical yarn with electronic features and then manufacture a textile material of that yarn.

It has been indicated in references (Hamdani et al., 2013; Kayacan and Yazgan Bulgun, 2009) that the resistance of heating panels, produced by weaving, is lower than that of knitted structures with the same dimensions, because of the structural characteristics: woven fabrics have inferior surface character as quality properties, when compared to knitted fabrics and heating elements made of non-woven materials. They have proved to be of little use owing to the high electrical resistance of conductive non-woven fabric. So, knits are more appropriate by implication, as a heating panel. There are two types of electrically conductive yarns – natural conductive (e.g. with conductive core) or with natural conductive additives (metal, carbon, etc.) and specially processed (coated with metals, metal salts or conductive polymers) of a common thread of textile fibres. Yarns that have electrical conductivity in their fibre composition by introducing natural conductive additives or special pre-treatment are more suitable for integration into textile materials than metal wire or cable since they do not limit textile properties (softness, flexibility) (Ohgushi et al., 1991; Tao, 2005). However, it should be noted that yarns obtained by conductive coatings covering their components in various ways tend to have a higher electrical resistance than metal (Locher, 2006). This needs to be evaluated when designing fabrics for specific conductive applications with intended electrical and thermal efficiency properties.

The creation of textile products with heating properties entails the manufacturing of a system based on a set of elements that must ensure the desired functionality of a particular fabric. Even if there are several kinds of textile heating elements, the research concentrates on functional products whose structures involve various electroconductive yarns specially designed for this special goal. Currently, such yarns characterize characteristics which are comparable to those of conventional textile yarns (fineness, flexibility, the ability to be processed on machines that are specific to the textile industry), due to their advanced development. As a result, the electroconductive yarns can be easily and effectively inserted into textile fabrics, thus allowing the use of a broad range of realization and manufacturing techniques that are unique to this field of activity.

There is quite an extensive supply of electrically conductive fibres and yarns made from them, suitable for the formation of heating contours in clothing products on the market. Acor (USA) manufactures fibres treated with pure silver (content ~ 15 %), named X-Static. Europa NCT Sp. (Poland) is the creator of new electrically conductive copper sulphide-coated polyacrylic fibres with a EURO-static label. Schoeller GmbH, Inc.CO KG (Austria) and Durafil (China) produce electrodes for a thread containing stainless steel fibres (5-95 %) under the name INOX. The stainless steel fibre, in this case, is thinner than a human hair, so the clothes made from the heterogeneous fibres do not irritate the skin. Bekaert (Belgium) manufactures BEKINOX® filament yarns of stainless steel. Tekstina (Slovenia) produces yarns with the name Tekstim™, which comes in the form of cotton fibres by also introducing stainless steel fibres. The R.Stat/N fibres produced by the company R.Stat (France) consist of PA 6.6 or polyester fibre coated with 0.2 μm electrolytic copper sulphide and a SilveR.Stat® silver layer. Electric conductors named Shieldex® PA and PES Fiber (non-insulated and TPU-coated) threads are coated with silver, copper, and also made of nickel by Statex Produktions- und Vertriebs GmbH (Germany). PA silver-plated yarn (unpolished and insulated TPU/PVC sheath) called Elitex® is manufactured by Imbut GmbH (Germany). Electrisola Feindraht AG (Switzerland) manufactures enamelled metal (copper, copper with silver coating, brass, silver, aluminium) 0.01–0.50 mm diameters of monofilament. The electrical conductive properties of the Swiss Shield® yarn manufactured by Swiss Shield AG (Switzerland) are secured to the textile yarn by incorporating a metal (copper, brass, bronze, gold, aluminium or steel) thread. W. Zimmermann GmbH & Co (Germany) manufactures cores for Novonic® and Textronic Inc. (India) Texro-Yarns® yarns have elastic properties. An extremely wide assortment of conductive yarns using various metals (stainless steel, aluminium, copper, nickel, tungsten, and iron) and their blended alloys are manufactured by Tibtex (France). These are various insulated and non-insulated yarn/brand names: Thermotech®, Copernic®, Silverpam®, Spuntex®, Tibtal®, Polynox®, Thermosew®. The company also manufactures Tibgrid® and Thermostretch® tapes for various intelligent textile heating parameters. Electrostatic materials and contours for heating textiles are also developed and manufactured by ITP GmbH, W. Zimmermann GmbH & Co., Gustav Gerster GmbH & Co. KG (Germany), Baltex (Feratec®, Great Britain) and others.

11.3. Applications and functions of smart textiles

Five functions can be distinguished in intelligent clothing (Das et. al., 2013; Chudasama, 2014; Van Langenhove and Hertleer, 2004):

- Sensors,
- Data processing
- Actuators,
- Storage,
- Intercommunication.
-

The basis of the *sensor* is to transform a signal into another signal; that can be read and understood by a predefined reader, which can be a device or a person. As for real devices, ultimately most signals are being transformed into electric ones (Das et al., 2013).

Textile materials cover a large area of the body and they are an excellent measuring tool. Biosignals that are: temperature, biopotentials (cardiogram, myography), acoustic (of heart, lungs, digestion, joints), ultrasound (blood flow), motion (respiration), humidity (sweat), pressure (blood) (Das et al., 2013; Chudasama, 2014).

Data processing is one of the components that are required only when active processing is necessary. Textile sensors could provide a huge amount of data, but problems arise due to large variations of signals between patients, complex analysis of stationary and time depending signals, lack of objective standard values, lack of understanding of complex interrelationships between parameters. As the textile material itself does not have any computing power, electronics, which can be miniaturized and flexible should be available, which leads to limited waterproof characteristics (Das et al., 2013; Chudasama, 2014).

Actuators respond to an impulse from the sensor function, possibly after data processing. Shape memory materials in the form of threads are the best-known examples of this area. They are able to react to a temperature change and can be used as an actuator and link up perfectly with the requirements imposed on smart textiles (Das et al., 2013; Chudasama, 2014).

Storage of data or energy is most common, sensing, data processing, actuation, communication; they usually need energy, mostly electrical power. Efficient energy management will consist of an appropriate combination of energy supply and energy storage capacity (Das et al., 2013).

For intelligence textiles, *intercommunication* has many faces: communication may be required within one element, between the individual elements to pass information. The communication is currently realized by either optical fibres or conductive yarns. Communication is possible by the following technologies: for the development of a flexible textile screen, the use of optical fibres (Das et al., 2013; Chudasama, 2014).

Post and Orth (1997) developed a fabric that can sense pressure. When the fabric is pressed at the right points, the two conducting layers make contact with nylon net and electric current flows from a row electrode to a column electrode. Light-emitting diodes (LED) with fuzzy conductive touch and close fasteners "Velcro" ends for electrical contacts are placed throughout the net. When both ends of an LED brush against the power and ground planes, the circuit is complete and the LED lights.

El-Sherif et al. (2007) developed a novel class of fibre optic chemical sensors, for detection of toxic and biological materials. The design of these fibre optic sensors was based on a cladding modification approach. The original passive cladding of the fibre, in a small section, was removed and the fibre core was coated with a chemically sensitive material. Any change in the optical properties of the modified cladding material, due to the presence of specific chemical vapour, changed the transmission properties of the fibre and resulted in modal power redistribution in multimode fibres. Both total intensity and modal power distribution (MPD) measurements were used to detect the output power change through the sensing fibres. The MPD method measures the power changes in the far field pattern, i.e. spatial intensity modulation in two dimensions. Conducting polymers, such as polyaniline and polypyrrole, experience a reversible change in conductivity upon exposure to chemical vapours. Conductivity change has been determined to be accompanied by a change in the optical properties of the material. Therefore, polyaniline and polypyrrole were selected as the modified cladding material for the detection of hydrochloride (HCl), ammonia (NH_3), hydrazine (H_4N_2), and dimethyl-methyl-phosphonate (DMMP) {a nerve agent, sarin stimulant}, respectively. Several sensors were prepared and successfully tested. The results showed obvious improvement in the sensor sensitivity when the MPD technique was used.

Innis et al. (2002) reported the development of complete textile-based batteries. Inherently conducting polymers are also being developed for wearable energy storage systems. These fabric batteries and photovoltaic ones have been developed utilizing ionic liquid electrolytes that could be incorporated into textile membranes. Fabric-based electrodes have been used to form a flexible and conformable battery system. These textile-based electrodes were formed by coating conducting polymeric materials on fabrics or by in situ or vapour phase polymerization of the monomers of conductive polymers on fabric substrates.

Huang et al. (2015) fabricated a weavable, knittable and wearable yarn supercapacitor from reduced-graphene-oxide-modified conductive yarns covered with a hierarchical structure of MnO_2 nanosheets and a polypyrrole thin film. The resultant modified yarns exhibited specific capacitances as high as 36.6 mF cm^{-1} and 486 mF cm^{-2} in aqueous electrolyte (three-electrode cell) or 31 mF cm^{-1} and 411 mF cm^{-2} in all solid-state two-electrode cells. The symmetric solid-state supercapacitor had high energy densities of 0.0092 mWh cm^{-2} and 1.1 mWh cm^{-3} (both normalized to the whole device) with long life cycle. Large energy storage textiles were produced by weaving developed flexible all-solid-state supercapacitor yarns to a 15×10 cm fabric on a loom and knitting in a woollen wrist band to form a pattern, enabling two functions: energy storage and wearability.

CONCLUSIONS

The significant attention that electroconductive textile materials have received recently can be attributed to the possibility of integrating the functionality of the material into a flexible textile structure or even into a garment. Insertion of conductive yarns into fabrics provides added value by decreasing the electrical resistance of fabrics and increasing shielding effects by reflection or absorption mechanisms of electromagnetic waves.

Due to the emergence of intelligent or smart textiles, a growing interest in electrically conductive textile materials has arisen. Their properties make it possible to add some functionality to the textile products in several ways:

- Providing electrical energy to a system;
- Transporting data (analogous or numerical);
- Making sensors (resistive or capacitive);
- Creating heat based on the Joule effect;
- Dissipating electrical charges (anti-static);
- Protection against electromagnetic interference (EMI).

According to the literature review, it is clear that the best EMI properties over the wide range of microwave wavelengths are exhibited by conductive fabrics that feature the lowest reflectance and transmission coefficients in addition to the greatest absorption factors. The textile materials, in which metallized yarns or coatings are incorporated, exhibit the highest shielding effectiveness in this frequency region until a maximum of 6 to 10 GHz is achieved. These fabrics are highly conductive and provide electromagnetic radiation (EMR) shields mostly via surface reflection. This can also be attributed to the presence of smaller wavelengths at higher frequencies.

Comparing the EMI shielding behaviour of coated and carbon-based fabrics, it can be stated that carbon-containing fabrics normally do not possess isotropic EM shielding behaviour due to the yarn direction obtained by weaving and knitting processes.

It should be noted that the colour change is significant in cases in which carbon treatment is applied to fabrics; depending on carbon quantity, the colour may turn from grey to black. The colour can also be a critical parameter for fabrics with metals because it is not always acceptable to have textiles with grey metal colour in an application.

Conductive polymers are the most promising EMR shielding materials among other conductive additives that are used to develop conducting textile materials with required properties intended for EMI shielding in the microwave range. Requirements for ideal EMI shielding textile materials, including fabrics coated with inherently conductive polymers (ICPs), present not only high EMR shielding effectiveness over a wide range of frequencies but also provide stable electrical properties, resistance to washing, and withstanding normal wear and tear.

As science goes forward, various and more cost-effective ways to increase the electrical conductivity of textiles using different polymer or chemical finishes appear. It is also likely that more application fields of conductive textiles will occur in the future and scientists will find a solution for conductive coating application on textiles that remain undamaged by washing.

REFERENCES

Abdelal, N., 2018. Electromagnetic interference shielding of stitched carbon fiber composites. Journal of Industrial Textiles, 1528083718798632.

Amiet, A., 2003. Free space permittivity and permeability measurements at microwave frequencies. Doctoral thesis. Monash University. 337.

Anderson, K. S. A. M., Seyam, A. M., 2004. The road to true wearable electronics. *TEXTILES MAGAZINE*, 31(1), 17-22.

Avloni, J.; Lau, R.; Ouyang, M.; Florio, L.; Henn, A. R.; Sparavigna, A., 2008. Polypyrrole-coated Nonwovens for Electromagnetic Shielding. Journal of industrial Textiles, 38(1), 55-68.

Avloni, J., Ouyang, M., Florio, L., Henn, A.R., Sparavigna, A., 2007. Shielding effectiveness evaluation of metallized and polypyrrole-coated fabrics. Journal of Thermoplastic Composite Materials, *20*(3), 241-254.

Bagavathi, M., 2015. A review on the production methods and testing of textiles for electromagnetic interference (EMI) shielding. International Journal of Engineering Research and Applications, 5(2), 34-40.

Bai, Y.: Li, H.: Gan, S.: Li, Y.: Liu, H.: Chen, L. 2018. Flexible heating fabrics with temperature perception based on fine copper wire and fusible interlining fabrics. Measurement, *122*, 192-200.

Banaszczyk, J.; De Mey, G.; Schwarz, A.; Van Langenhove, L. 2009. Current Distribution Modelling in Electroconductive Fabrics. Fibres & Textiles in Eastern Europe, 17, 2(73), 28-33.

Berberi, P. G. 1998. A New Method for Evaluating Electrical Resistivity of Textile Assemblies. Textile Research Journal, *68*(6), 407-412.

Berberi, P. G. 2001. Effect of processing on electrical resistivity of textile fibers. Journal of Electrostatics, *51*, 538-544.

Brazis, R.; Czekalski, J.; Kozakiewicz, D.; Michalak, M.; Stasiak, M. 2000. Electromagnetic Wave Attenuation in Hemp-PAN Fibre Blends. Fibres & Textiles in Easter Europe, *8*(2), 35-38.

Brzeziński, S.; Rybicki, T.; Malinowska, G.; Karbownik, I.; Rybicki, E.; Szugajew, L. 2009. Effectiveness of shielding electromagnetic radiation, and assumptions for designing the multi-layer structures of textile shielding materials. Fibres & Textiles in Eastern Europe, 1(72), 60-65.

Celozzi, S.; Araneo, R.; Lovat, G. 2008. Electromagnetic shielding. Hoboken, NJ: John Wiley & Sons Inc., 353.

CEN/TR 16298: 2011. Textiles and textile products – Smart textiles – Definitions, categorisation, applications and standardization needs.

Chen H. C.; Lee K. C.; Lin, J. H. 2004. Electromagnetic and Electrostatic Shielding Properties of Co-weaving-knitting Reinforced Composites. Composites: Part A, *35*, 1249-1256.

Chen, H. C.; Lee, K. C.; Lin, J. H.; Koch, M. 2007. Comparison of electromagnetic shielding effectiveness properties of diverse conductive textiles via various measurement techniques. Journal of Materials Processing Technology, *192*, 549-554.

Chen, H.C.; Lin, J.H.; Lee, K.C. 2008. Electromagnetic shielding effectiveness of copper/stainless steel/polyamide fiber co-woven-knitted fabric reinforced polypropylene composites. Journal of Reinforced Plastics and Composites, *27*(2), 187-204.

Cheng, K. B. 2000. Production and electromagnetic shielding effectiveness of the knitted stainless steel/polyester fabrics. Journal of Textile Engineering, 46(2), 42-52.

Cheng, K. B.; Cheng, T. W.; Nadaraj, R. N.; Dev, V. G.; Neelakandan, R. 2006. Electromagnetic shielding effectiveness of the twill copper woven fabrics. Journal of Reinforced Plastics and Composites, *25*(7), 699-709.

Cheng, K.B.; Lee, M.L.; Ramakrishna, S.; Ueng, T.H. 2001. Electromagnetic shielding effectiveness of stainless steel/polyester woven fabrics. Textile Research Journal, *71*(1), 42-49.

Cheng, K. B.; Ramakrishna, S; Lee, K. C. 2000. Development of conductive knitted-fabric-reinforced thermoplastic composites for electromagnetic shielding applications. Journal of Thermoplastic Composite Materials, *13*(5), 378-399.

Chubb, J. N. 2004. Comments on methods for charge decay measurement. Journal of Electrostatics, *62*(1), 73-80.

Chudasama, D. 2014. Wearable electronics. International Journal of Electrical and Electronics Engineering Research (IJEEER), 4(5), 27-32

Cisło, R.; Krucińska, I.; Babeł, K.; Jurewicz, K. 2003. Resistance of carbon and active carbon precursor nonwovens. Fibres & Textiles in Eastern Europe, 2(41), 75-78.

Cisło, R.; Krucińska, I.; Babeł, K.; Koszewska, M. 2004. Electrical resistivity of active carbon nonwovens. Fibres & Textiles in Eastern Europe, 3(47), 70-74.

Colaneri, N. F.; Schacklette, L. W. 1992. EMI shielding measurements of conductive polymer blends. IEEE transactions on instrumentation and measurement, 41(2), 291-297.

Dall'Acqua, L.; Tonin, C.; Peila, R.; Ferrero, F.; Catellani, M. 2004. Performances and properties of intrinsic conductive cellulose–polypyrrole textiles. Synthetic Metals, *146*(2), 213-221.

Das, N. C.; Khastgir, D.; Chaki, T. K.; Chakraborty, A. 2002. Electromagnetic interference shielding effectiveness of hybrid conductive polymer composite. Journal of Elastomers & Plastics, 34(3), 199-223.

Das, S. C.; Paul, D.; Hasan Sk. Md. M.; Chowdhury, N.; Nizam, Md. E. H. 2013. Smart textiles – New possibilities in textile engineering. Internal conference on Mechanical, industrial and materials engineering (ICMIME2013), RUET, Rajshahi, Bangladesh. Paper ID: IE-31, 514-519.

Davydov, B. E.; Tkachenko, A. V.; Maureen, B. P.; Tsarevskiy, B. 1972. Method for Determining the Electrical Resistance of Fibrous Materials. Chemical Fibers, 3, 71-72 (in Russian).

De Mey, G.; ÖZÇELİK, M.; Schwarz, A.; Kazani, I.; Hertleer, C.; Van Langenhove, L.; GÜRSOY N. Ç. 2014. Designing of conductive yarn knitted thermal comfortable shirt using battery operated heating system. Journal of Textile and Apparel/Tekstil ve Konfeksiyon, 24.1, 26-29

De Seze, R.; Mausset, A. L.; Ayoub, J.; Pina, G.; Miro, L.. 2001. Evaluation of the health impact of the radio-frequency fields from mobile telephones. Indoor and Built Environment, *10*(5), 284-290.

De Temmerman, L. 1992. New metallized materials for EMI/RFI shielding. Journal of Coated Fabrics, *21*(3), 191-198.

Ding, J. T. F.; Tao, X.; Au, W. M.; Li L. 2014. Temperature effect on the conductivity of knitted fabrics embedded with conducting yarns. Textile Research Journal, *84*(17), 1849-1857.

Dhawan, A.; Seyam, A. M.; Ghosh T. K.; Muth, J. F. 2004. Woven Fabric-Based Electrical Circuits: Part I: Evaluating Interconnect Methods. Textile Research Journal, 74(10), 913-919.

Duran, D.; Kadoğlu, H. 2015. Electromagnetic shielding characterization of conductive woven fabrics produced with silver-containing yarns. Textile Research Journal, 85(10), 1009-1021.

El-Newashy, R. F.; Saad, M. A.; Turky, G. M.; Dokki, E. B. S. 2012. Integration of conductive yarns into fabric by stitching. Journal of American Science, 8(3), 213-217.

El-Sherif, M.; Bansal, L.; Yuan, J. 2007. Fiber optic sensors for detection of toxic and biological threats. Sensors, *7*(12), 3100-3118.

EN 1149-1: 2000. Protective clothing - Electrostatic properties - Part 1: Test method for measurement of surface resistivity (withdrawn).

EN 1149-1: 2006. Protective clothing - Electrostatic properties - Part 1: Test method for measurement of surface resistivity

EN 1149-2: 2000. Electrostatic properties of protective clothing - Part 2: Test method for measurement of the electrical resistance through a material (vertical resistance)

EN 1149-3: 2004. Protective clothing - Electrostatic properties - Part 3: Test methods for measurement of charge decay

EN 16812:2016. Textiles and textile products - Electrically conductive textiles - Determination of the linear electrical resistance of conductive tracks

Engin, F. Z.; Usta, İ. 2015. Development and characterisation of polyaniline/polyamide (PANI/PA) fabrics for electromagnetic shielding. The Journal of The Textile Institute, 106(8), 872-879.

EN ISO 11611:2015. Protective clothing for use in welding and allied processes (ISO 11611:2015).

Fu, K. K.; Padbury, R.; Toprakci, O.; Dirican, M.; Zhang, X. 2017. Conductive textiles. Engineering of High-Performance Textiles, 13, 305.

Foitzik, R. C.; Kaynak, A.; Pfeffer, F. M. 2007. Conductive poly (α, ω-bis (3-pyrrolyl) alkanes)-coated wool fabrics. Synthetic metals, *157*(13-15), 534-539.

Gasana, E.; Westbroek, P.; Hakuzimana, J.; De Clerck, K.; Priniotakis, G.; Kiekens, P.; Tseles, D. 2006. Electroconductive textile structures through electroless deposition of polypyrrole and copper at polyaramide surfaces. Surface and Coatings Technology, 201(6), 3547-3551.

Geetha, S.; Kumar, K. K. S.; Trivedi, D. C. 2005. Conducting fabric-reinforced polyaniline film using p-chlorophenol as secondary dopant for the control of electromagnetic radiations. Journal of composite materials, *39*(7), 647-658.

Geetha, S.; Satheesh Kumar, K. K.; Rao, C. R.; Vijayan, M.; Trivedi, D. C. 2009. EMI shielding: Methods and materials - A review. Journal of applied polymer science, *112*(4), 2073-2086.

Ghosh, T.; Dhawan, A. 2006. Electronic Textiles and Their Potential. Indian Journal of Fibre & Textile Research, 31(1), 170-176.

Gonzalez, J. A. 2005. Textiles for protection. Woodhead publishing limited, 754.

Gonzalez, J. A.; Rizvi, S. A.; Crown, E. M.; Smy, P. R. 2001. A Laboratory Protocol to Assess the Electrostatic Propensity of Protective-clothing Systems. Journal of the Textile Institute, *92*(3), 315-327.

Guanghui, W.; Shanghe, L.; Wei, T.; Yazhou, C. 1998. A new test method and device of frictional voltage decaying for fabrics. Journal of electrostatics, *44*(1-2), 105-111.

Gupta, K.; Abbas S.; Abhyankar, A. 2016. Carbon black/polyurethane nanocomposite-coated fabric for microwave attenuation in X & Ku-band (8–18 GHz) frequency range. Journal of Industrial Textiles, 46(2), 510–529.

Håkansson, E.; Amiet, A.; Kaynak, A. 2007. Dielectric characterization of conducting textiles using free space transmission measurements: Accuracy and methods for improvement. Synthetic Metals, 157(24), 1054-1063.

Hamdani, S. T. A.; Potluri, P.; Fernando A.; 2013, Hamdani, S.; Potluri, P.; Fernando, A. 2013. Thermo-mechanical behavior of textile heating fabric based on silver coated polymeric yarn. Materials, *6*(3), 1072-1089.

Hersh, S. P.; Montgomery, D. J. 1952. Electrical resistance measurements on fibers and fiber assemblies. Textile Research Journal, *22*(12), 805-818.

Hersh, S. P.; Montgomery, D. J. 1955. Static electrification of filaments: experimental techniques and results. Textile Research Journal, *25*(4), 279-295.

Hersh, S. P.; Montgomery, D. J. 1956. Static Electrification of Filaments Theoretical Aspects. Textile Research Journal, *26*(12), 903-913.

Hoback, J. T.; Reilly, J. J. 1988. Conductive coatings for EMI shielding. Journal of Elastomers & Plastics, *20*(1), 54-69.

Holme, I.; McIntyre, J. E.; Shen, Z. J. 1998. Electrostatic charging of textiles. Textile Progress, 28(1), 1–85.

Huang, Y.; Hu, H.; Huang, Y.; Zhu, M.; Meng, W.; Liu, C.; Pei, Z.; Hao, C.; Wang, Z.; Zhi, C. 2015. From industrially weavable and knittable highly conductive yarns to large wearable energy storage textiles. ACS nano, *9*(5), 4766-4775.

IEC 61340-5-1:2016 Electrostatics - Part 5-1: Protection of electronic devices from electrostatic phenomena - General requirements.

Innis, P.C.; Mazurkiexicz, J.; Aboutanos, V.; Ding, J.; Spinks, G.M.; Zhou, D.; Wu, J.; Wallace, G.G.; MacFakane, D.; Macfallane, D.; Forsyth, M.; Forsyth, S. Inherently conducting polymers for wearable energy conversion and storage systems. Paper presented in

International Interactive Textiles for The Warrior Conference, Cambridge, MA, 9-11, July, 2002.
Jachowicz, M. 2013. Electrostatic properties of selected personal protective equipment regarding explosion hazard. Journal of Sustainable Mining, 12(1), 27-33.
Järvinen, K.; Puolakka, A. 2003. Weathering of polyaniline-treated polyester filter fabrics. Textile Research Journal, *73*(7), 593-596.
Jonassen, N. 2013. Electrostatics. Springer Science & Business Media, 196
Joseph, N.; Varghese, J.; Sebastian, M. T. 2017. In situ polymerized polyaniline nanofiber-based functional cotton and nylon fabrics as millimeter-wave absorbers. Polymer Journal, 49(4), 391.
Kacprzyk, R.; Mista, W. 2006. Back corona in fabrics. Fibres & Textiles in Eastern Europe, *14*(5), 59, 35-38.
Kalliohaka, T.; Paasi, J.; Lemaire, P. 2005. Forced corona method for the evaluation of fabrics with conductive fibres. Journal of electrostatics, *63*(6-10), 583-588.
Kan, C. W.; Yuen, C. W. M. 2008. Static properties and moisture content properties of polyester fabrics modified by plasma treatment and chemical finishing. Nuclear Instruments and Methods in Physics Research Section B: Beam Interactions with Materials and Atoms, *266*(1), 127-132.
Kathirgamanathan, P.; Toohey, M. J.; Haase, J.; Holdstock, P.; Laperre, J.; Schmeer-Lioe, G. 2000. Measurements of incendivity of electrostatic discharges from textiles used in personal protective clothing. Journal of Electrostatics, *49*(1-2), 51-70.
Kayacan, O.; Yazgan Bulgun, E. 2009. Heating behaviors of metallic textile structures. International Journal of Clothing Science and Technology, *21*(2/3), 127-136.
Keith, J. M.; Janda, N. B.; King, J. A.; Perger, W. F.; Oxby, T. J. 2005. Shielding effectiveness density theory for carbon fiber/nylon 6, 6 composites. Polymer composites 26(5), 671-678.
Kessler, L.; Fisher, W. K. 1997. A study of the electrostatic behavior of carpets containing conductive yarns. Journal of electrostatics, *39*(4), 253-275.
Kim, S. H.; Jang, S. H.; Byun, S. W.; Lee, J. Y.; Joo, J. S.; Jeong, S. H.; Park, M. J. 2003. Electrical properties and EMI shielding characteristics of polypyrrole–nylon 6 composite fabrics. Journal of Applied Polymer Science, 87(12), 1969-1974.

Kim, B.; Koncar, V.; Devaux, E.; Dufour, C.; Viallier, P. 2004. Electrical and morphological properties of PP and PET conductive polymer fibers. Synthetic Metals, *146*(2), 167-174.

King, J. A.; Pisani, W. A.; Klimek-McDonald, D. R.; Perger, W. F.; Odegard, G. M.; Turpeinen, D. G. 2015. Shielding effectiveness of carbon-filled polypropylene composites. Journal of Composite Materials, 50(16), 2177-2189.

King, J. A.; Weber, E. H.; Quan, C.; Tucker, K. W.; Vogt, B. D. 2000. Conductive high temperature nylon. Journal of composite materials, *34*(24), 2038-2060.

Knittel, D.; Schollmeyer, E. 2009. Electrically high-conductive textiles. Synthetic Metals, *159*(14), 1433-1437.

Koprowska, J.; Pietranik, M.; Stawski, W. 2004. New type of textiles with shielding properties. Fibres & Textiles in Eastern Europe, 3(47), 39-42.

Koralewski H. E. 2006, Energiehaushalt und Temperaturregulation. Berlin, Charité Bioinformatic SS Cluster B4, 1. Access in internet: https://klinphys.charite.de/bioinfo/2_p-skripten/b4_b_waermehaushalt.pdf.

Kowalski, J. M.; Wroblewska, M. 2006. Premises for Practical Evaluation of the Anti-electrostatic Properties of Protective Garments. Fibres & Textiles in Eastern Europe, *5*(59), 23-28.

Kuhn, H. H. 1997. Adsorption at the Liquid/Solid Interface: Conductive Textiles Based on Polypyrrole. Textile Chemist & Colorist, 29(12), 17-21.

Lai, A. C. 2006. Investigation of electrostatic forces on particle deposition in a test chamber. Indoor and Built Environment, 15(2), 179-186.

Lai, K.; Sun, R. J.; Chen, M. Y.; Wu, H.; Zha, A. X. 2007. Electromagnetic shielding effectiveness of fabrics with metallized polyester filaments. Textile Research Journal, *77*(4), 242-246.

Lambrozo, J. 2001. Electric and Magnetic Fields with a Frequency of 50–60 Hz: Assessment of 20 Years of Research. Indoor and Built Environment, 10(5), 299-305.

Lee Sandbach, D.; Burkitt, J.; Walkington, S. M.; Crispin, P. G. 2008. Knitted sensor. United States Patent Application Publication–2008.-Nr. US, 7377133.

Lei, C.H.; Lin, J.H.; Chang, C.H.; Lou, C. W. 2004. The residual electrostatic voltage of laminated nonwovens with different surface patterns. Textile Research Journal, *74*(5), 383-386.

Lerner, A. 1985. A new additive for electrostatic discharge control in foams and elastomers. Journal of cellular plastics, *21*(1), 31-34.

Lin, J.H.; Lou, C. W. 2003. Electrical properties of laminates made from a new fabric with PP/stainless steel commingled yarn. Textile Research Journal, *73*(4), 322-326.

Locher I. 2006. Technologies for System-on-Textile Integration, Doctoral thesis, ETH No. 16467, Swiss Federal Institute of Technology: Zürich, p. 121, DOI 10.3929/ethz-a-005135763

Lutz, B.; Kindersberger, J. 2009. Influence of relative humidity on surface charge decay on epoxy resin insulators. In 2009 IEEE 9th International Conference on the Properties and Applications of Dielectric Materials, 883-886.

Maity, S.; Chatterjee, A. 2018. Conductive polymer-based electro-conductive textile composites for electromagnetic interference shielding: A review. Journal of Industrial Textiles, *47*(8), 2228-2252.

Maity, S.; Singha, K.; Debnath, P.; Singha, M. 2013. Textiles in electromagnetic radiation protection. Journal of Safety Engineering, *2*(2), 11-19.

Marchini, F. 1991. Advanced applications of metallized fibres for electrostatic discharge and radiation shielding. Journal of Coated Fabrics, *20*(3), 153-166.

Matukonis, A.; Palaima, J.; Vitkauskas, A. 1976. Tekstilės medžiagotyra, publishing house Mokslas, 424 (in Lithuanian).

Mečnika, V.; Hoerr, M.; Krivinš, I.; Schwarz, A. 2014. Smart textiles for healthcare: applications and technologies. In Rural Environment. Education. Personality (REEP) Proceedings of the International Scientific Conference (Latvia). Latvia University of Agriculture. 150-161

Mehmood, T.; Dai, X. J.; Kaynak, A.; Kouzani, A. 2012. Improved bonding and conductivity of polypyrrole on polyester by gaseous plasma treatment. Plasma processes and polymers, 9(10), 1006-1014.

Michalak, M.; Bilska, J. 2002.'Research into manufacturing nonwovens with barrier properties'. In Proceedings of the 7th International Conference on the Science and the Quality of Life, Vilnius, Lithuania, June 2002, Studium Vilnense,. 323-326.

Michalak, M.; Brazis, R.; Kazakevicius, F.; Bilska, J.; Krucińska, I. 2006. Nonwovens with implanted split rings for barriers against electromagnetic radiation. Fibres & Textiles in Eastern Europe, 5(59), 64-68.

Michalak, M.; Kazakevičius, V.; Dudzińska, S.; Krucińska, I.; Brazis, R. 2009. Textiles embroidered with split-rings as barriers against microwave radiation. Fibres & Textiles in Eastern Europe, 1(72), 66-70.

Michalak, M.; Krucińska, I.; Brazis, R.; Vitkauskas, A.; Skrzetuska, E. 2006. Studies on multilayer nonwovens containing flax fibres designed for electromagnetic radiation barriers. Materials Science (Medžiagotyra), 12(1), 69-72.

Molina, J.; Del Rio, A.I.; Bonastre, J.; Cases, F. 2009. Electrochemical polymerisation of aniline on conducting textiles of polyester covered with polypyrrole/AQSA. European Polymer Journal, *45*(4), 1302-1315.

Montgomery, D. J.; Smith, A. E.; Wintermute, E. H. 1961. Static electrification of filaments: effect of filament diameter. Textile Research Journal, 31(1), 25-31.

Motyl, E.; Lowkis, B. 2006. Effect of air humidity on charge decay and lifetime of PP electret nonwovens. Fibres and Textiles in Eastern Europe, *14*(5), 59, 39-42.

Nakamura, K.; Hatakeyama, T.; Hatakeyama, H. 1981. Studies on bound water of cellulose by differential scanning calorimetry. Textile Research Journal, *51*(9), 607-613.

Nakamura, K.; Hatakeyama, T.; Hatakeyama, H. 1983. Effect of bound water on tensile properties of native cellulose. Textile Research Journal, *53*(11), 682-688.

Nayak, L.; Khastgir, D.; Chaki, T. K. 2013. A mechanistic study on electromagnetic shielding effectiveness of polysulfone/carbon nanofibers nanocomposites. Journal of Materials Science, 48(4), 1492-1502.

Neha, A.; Sivasubramanian, S. P.; Vanbhatte, S. B.; Kadole, P. V. 2013. Polypyrrole/multiwall carbon nanotube/viscose composites for smart fabrics. International Journal of engineering research & technology, 2(2), 1-11.

Nelson, M. A.; Rogers, R. L.; Gilmartin, B. P. 1993. Anti-static mechanisms associated with FIBC fabrics containing conductive fibres. Journal of Electrostatics, 30, 135-147.

Nurmi, S.; Hammi, T.; Demoulin, B. 2007. Protection against electrostatic and electromagnetic phenomena. In Multifunctional Barriers for Flexible Structure, Springer, Berlin, Heidelberg, 63-83.

Ohgushi, K.; Hijiri, M.; Kitazawa, Z. 1991. U.S. Patent No. 4,983,814. Washington, DC: U.S. Patent and Trademark Office.

Ono, H.; Ohsawa, A.; Tabata, Y. 2003. New method for evaluating antistatic effect in floor coverings. Journal of electrostatics, 57(3-4), 355-362.

Onogi, Y.; Sugiura, N.; Matsuda, C. 1997. Temperature effect on dissipation of triboelectric charge into air from textile surfaces. Textile research journal, *67*(1), 45-49.

Onogi, Y.; Sugiura, N.; Nakaoka, Y. 1996. Dissipation of triboelectric charge into air from textile surfaces. Textile research journal, *66*(5), 337-342.

Ouyang, Y.; Chappell, W. 2005. Characterization of the High Frequency Properties of Electrotextiles. Proceedings of 28th General Assembly of International Union of Radio Science (URSI), vol. 1, 1106- 1108.

Paasi, J.; Fast, L.; Kalliohaka, T.; Smallwood, J.; Börjesson, A.; Haase, J.; Vogel, C.; Lemaire, P.; Coletti, G.; Guastavino, F.; Peltoniemi, T.; Reina, G. 2005. Risks of damage to electronics with reference to charged clothing. Journal of electrostatics, 63(6-10), 603-608.

Paasi, J.; Kalliohaka, T.; Luoma, T.; Soininen, M.; Salmela, H.; Nurmi, S.; Coletti, G.; Guastavino, F.; Fast, L.; Nilsson, A.; Lemaire, P.; Laperre, J.; Vogel, C.; Haase, J.; Peltoniemi, T.; Viheriäkoski, T.; Reina, G.; Smallwood, J.; Börjesson A. 2004. Research Report NO BTUO45-041224. ESTAT Garments evaluation of existing test methods for ESD garments.

Paasi, J.; Nurmi, S.; Vuorinen, R.; Strengell, S.; Maijala, P. 2001. Performance of ESD protective materials at low relative humidity. Journal of Electrostatics, 51, 429-434.

Pan N.; Gibson P. 2006, Thermal and moisture transport in fibrous materials. Woodhead Publishing Series in Textiles No. 56, Cambridge, England, p. 632, ISBN: 9781845690571

Perlman, M. M. and Reedyk, C. W. 1968. Production and charge decay of film electrets. Journal of The Electrochemical Society, *115*(1), 45-49.

Perumalraj, R.; Dasaradan, B. S.; Anbarasu, R.; Arokiaraj, P.; Harish, S. L. 2009. Electromagnetic shielding effectiveness of copper core-woven fabrics. The Journal of The Textile Institute, 100(6), 512-524.

Petcu, I.; Agrawal, P.; Curteza, A.; Visser, R.; Brinks, G.; Teodorescu, M. 2012. A comparative study of heating elements used for the development of textile heaters. In 12th AUTEX World Textile Conference, Faculty of Textile Technology of the University of Zagreb.

Pietranik, M.; Stawski, W. 2004. New Type of Textiles with Shielding Properties. Fibres & Textiles in Eastern Europe, *12*(3), 47. 39-42.

Pinar, A.; Michalak, L. 2006. Influence of structural parameters of wale-knitted fabrics on their electrostatic properties. Fibres & Textiles in Eastern Europe, *14(*5), 69-74.

Poboroniuc, M. S.; Curteza, A.; Cretu, V.; Macovei, L. 2014. Designing wearable textile structures with embedded conductive yarns and testing their heating properties. In International Conference and Exposition on Electrical and Power Engineering (EPE), 778. DOI

10.1109/ICEPE.2014.6970016
Post, E. R.; Orth, M. 1997. Smart fabric, or washable computing. In 1st IEEE International Symposium on Wearable Computing (pp. 167-168).
Raji, R. K.; Miao, X.; Boakye, A. 2017. Electrical Conductivity in Textile Fibers and Yarns. AATCC Journal of Research, 4(3), 8-21.
Rakshit, A. K.; Hira, M. A. 2014. Electrically conductive fibre substrates. International Journal of Fiber and Textile Research, 4, 44-48.
Ramakrishnan, S. 2011. From a laboratory curiosity to the market place. Resonance, *16*(12), 1254-1265.
Ramer, E.M.; Richards, H. R.1968. Correlation of the electrical resistivities of fabrics with their ability to develop and to hold electrostatic charges. Textile Research Journal, *38*(1), 28-35.
Redlich, G.; Obersztyn, E.; Olejnik, M.; Fortuniak, K.; Bartczak, A.; Szugajew, L.; Jarzemski, J. 2014. New textiles designed for anti-radar camouflage. Fibres & Textiles in Eastern Europe, 1(103), 34-43.
Rivas, B. L.; Sanchez, C. O. 2001. Synthesis, characterization, and electrical conductivity of polyaniline derivatives: Study with the metal ions Cu (II), Ni (II), and Co (II). Journal of Applied Polymer Science, 82(2), 330-337.
Rizvi, S. A. H.; Crown, E. M.; Gonzalez, J. A.; Smy, P. R. 1998. Electrostatic characteristics of thermal-protective garment systems at various low humidities. The Journal of The Textile Institute, 89(4), 703-710.
Rizvi, S. A. H.; Crown, E. M.; Osei-Ntiri, K.; Smy, P. R.; & Gonzalez, J. A. 1995. Electrostatic characteristics of thermal-protective garments at low humidity. Journal of The Textile Institute, *86*(4), 549-558.
Roell, F.; 1996, U.S. Patent No. 5,484,983. Washington, DC: U.S. Patent and Trademark Office.
Roh, J. S.; Chi, Y. S.; Kang, T. J.; Nam, S. W.; 2008. Electromagnetic shielding effectiveness of multifunctional metal composite fabrics. Textile Research Journal, *78*(9), 825-835.
Rubežienė, V.; Abraitienė, A.; Baltušnikaitė-Guzaitienė, J.; Varnaitė-Žuravliova, S.; Sankauskaitė, A.; Kancleris, Ž.; Ragulis, P.; Šlekas, G. 2018. The influence of distribution and deposit of conductive coating on shielding effectiveness of textiles. The Journal of The Textile Institute, 109(3), 358-367.
Rubežienė, V.; Baltušnikaitė, J.; Varnaitė-Žuravliova, S.; Sankauskaitė, A.; Abraitienė, A.; Matuzas, J. 2015. Development and investigation of electromagnetic shielding fabrics with different electrically conductive additives. Journal of Electrostatics, 75, 90-98.

Rupprecht, L. 1999. Conductive Polymers and Plastics – In Industrial Applications, USA Norwich, NY: Plastic Design Library, 143.

Saini, P.; Arora, M. 2012. Microwave absorption and EMI shielding behavior of nanocomposites based on intrinsically conducting polymers, graphene and carbon nanotubes. In New Polymers for Special Applications. IntechOpen.

SEÇKİN UĞURLU, Ş.; ÖZKURT, A.; ÖZDEMİR. H. 2015. Comparison of methods for evaluating the shielding effectiveness of clothing woven fabrics. 6th International Technical Textiles Congress, 14-16 October, İzmir-TÜRKİYE, p. 1-7

Sereda, P. J.; Feldman, R. F. 1964. 27—Electrostatic charging on fabrics at various humidities. Journal of the Textile Institute Transactions, *55*(5), T288-T298.

Sezgin, H.; Bahadir, S. K.; Boke, Y. E.; Kalaoglu F. 2012, Effect of different conductive yarns on heating behaviour of fabrics. *In PMUTP international conference: textile & fashion*, Access to internet: http://textileconference.rmutp.ac.th/wp-content/uploads/2012/10/009-Effect-of-Different-Conductive-Yarns-on-Heating-Behaviour-of-Fabrics.pdf

Shahidi, S.; Moazzenchi, B. 2018. Carbon nanotube and its applications in textile industry–A review. The Journal of The Textile Institute, 1-14.

Shashoua, V. E. 1958. Static electricity in polymers. I. Theory and measurement. Journal of Polymer Science, *33*(126), 65-85.

Shateri-Khalilabad, M.; Yazdanshenas, M. E. 2013. Fabricating electroconductive cotton textiles using graphene. Carbohydrate polymers, 96(1), 190-195.

Singh, K.; Ohlan, A.; Dhawan, S. K. 2012. Polymer-graphene nanocomposites: preparation, characterization, properties, and applications. In Nanocomposites-New Trends and Developments. IntechOpen.

Stankutė, R.; Grinevičiūtė, D.; Gutauskas, M.; Žebrauskas, S.; Varnaitė, S. 2010. Evaluation of electrostatic properties of fiber-forming polymers. Materials Science (Medžiagotyra), *16*(1), 72-75.

Stoppa, M.; Chiolerio, A. 2014. Wearable electronics and smart textiles: a critical review. Sensors, *14*(7), 11957-11992.

Strazdienė, E.; Dobilaite, V.; 2007. Techninės Tekstilės gaminiai ir protingoji apranga. Šiauliai: VŠĮ Šiaulių universiteto leidykla, Lithuania, 168, (in Lithuanian) DOI 10.5755/e01.9786090204634

Su, C. I.; Chern, J. T. 2004. Effect of stainless steel-containing fabrics on electromagnetic shielding effectiveness. Textile Research Journal, *74*(1), 51-54.

Sweet, F. H.; Knobel, T. M.; Walker, M. K.; Thompson, C. P. 1986. Protection of Electrostatic Discharge (ESD) Sensitive Devices with Urethane Foams. Journal of cellular plastics, *22*(2), 139-146.

Šafářová, V.; Militký, J. 2017. Multifunctional metal composite textile shields against electromagnetic radiation—effect of various parameters on electromagnetic shielding effectiveness. Polymer Composites, *38*(2), 309-323.

Šafářová, V.; Tunák, M.; Militký, J. 2015. Prediction of hybrid woven fabric electromagnetic shielding effectiveness. Textile Research Journal, 85(7), 673-686.

Šahta, I.; Baltina, I.; Truskovska, N.; Blums, J.; Deksnis E. 2014, Selection of conductive yarns for knitting an electrical heating element. High Performance and Optimum Design of Structures and Materials, 137, 91-102.

Tanahashi, I.; Yoshida, A.; Nishino, A. 1990. Activated carbon fiber sheets as polarizable electrodes of electric double layer capacitors. Carbon, 28(4), 477-482.

Tao, X. 2005. Wearable Electronics and Photonics, Woodhead Publishing Limied, England, 256.

Tao, Y.; Li, P.; Shi, S. Q. 2016. Effects of Carbonization Temperature and Component Ratio on Electromagnetic Interference Shielding Effectiveness of Woodceramics. Materials, *9*(7), 540.

Tappura, K.; Nurmi, S. 2003. Computational modelling of charge dissipation of fabrics containing conductive fibres. Journal of Electrostatics, 58(1-2), 117-133.

Telipan, G.; Morari, C.; Moaşa, B. 2017. Electromagnetic shielding characterization of several conductive textiles. Bulletin of the Transylvania University of Brasov. Engineering Sciences. Series I, 10(1), 17-24

Tong, X. C. 2016. Advanced materials and design for electromagnetic interference shielding. CRC press, 344.

Van Langenhove, L.; Hertleer, C. 2004. Smart clothing: a new life. International journal of clothing science and technology, 16(1/2), 63-72.

Van Langenhove, L.; Puers, R.; Matthys, D. 2005. Intelligent textiles for protection. Textiles for protection, 176-195.

Vargas S. C. 2009. Smart Clothes - Textilien mit Elektronik. Was bietet der Markt der Intelligenten Bekleidung? Hamburg Diplomica® Verlag GmbH, 314, ISBN: 978-3-8366-7230-6

Varnaitė-Žuraliova, S.; Baltušnikaitė-Guzaitienė, J.; Valasevięiūtė, L.; Verbienė, R.; Abraitienė, A. 2016. Assessment of electrical characteristics of conductive woven fabrics. American Journal of Mechanical and Industrial Engineering. 1(3), 38-49

Vassiliadis, S.; Provatidis, C.; Prekas, K.; Rangoussi, M. 2004. Electrically Conductive Spun Yarns. In Proc. of the Xth International Izmir Textile and Apparel Symposium (Vol. 37), 49.

Vogel, C.; Beier, H.; Erth, H. 2006. Research and development in the field of special protective clothing; requirements for selected methods and products. Fibres & Textiles in Eastern Europe, 5(59), 29-34.

Von Pidoll, U. 2009. An overview of standards concerning unwanted electrostatic discharges. Journal of Electrostatics, *67*(2-3), 445-452.

Wainwright, H. L. 2016. Design, evaluation, and applications of electronic textiles. In Performance Testing of Textiles. Woodhead Publishing, 193-213.

Welsher, T. L.; Blondin, T. J.; Dangelmayer, G. T.; Smooha, Y. 1990. Design for electrostatic-discharge (ESD) protection in telecommunications products. AT&T Technical Journal, *69*, 77-96.

White, D. R.; Mardiguian, M. 1988. A Handbook Series on Electromagnetic Interference and Compatibility: EMI Control Methodology and Procedures. Interference Control Technologies.

Wilson, D.1963. 10—The electrical resistance of textile materials as a measure of their anti-static properties. Journal of the Textile Institute Transactions, 54(3), T97-T105.

www.acor.com

www.bekaert.com

www.elektrisola.com

www.europa.nct.pl

www.idtechex.com/research/reports/e-textiles-2018-2028-technologies-markets-and-players-000613.asp?viewopt=showall

www.imbut.de

www.r-stat.com

www.schoeller-wool.com

www.statex.de

www.swicofill.com

www.swiss-shield.ch

www.tekstina.si

www.tibtech.com

www.transparencymarketresearch.com/smart-fabrics-and-interactive-textiles-market.html

www.zimsi.com

www.textronicsinc.com

Xiao, H.; Shi, M. W.; Wang, Q.; Liu, Q. 2014. The electromagnetic shielding and reflective properties of electromagnetic textiles with pores, planar periodic units and space structures. Textile Research Journal, 84(16), 1679-1691.
Yuping, D.; Shunhua, L.; Hongtao, G. 2006. Investigation of electromagnetic characteristics of polyaniline composites. Journal of composite materials, 40(12), 1093-1104.
Zhang, X. 2011. Anti-static and conductive textiles. In Functional Textiles for Improved Performance, Protection and Health, Woodhead Publishing.; 27-44.
Zhang, H.; Tao, X.; Wang, S.; Yu, T. 2005. Electro-mechanical properties of knitted fabric made from conductive multi-filament yarn under unidirectional extension. Textile research journal, *75*(8), 598-606.
Zhong, S.; Padeste, C.; Kazacos, M.; Skyllas-Kazacos, M. 1993. Comparison of the physical, chemical and electrochemical properties of rayon-and polyacrylonitrile-based graphite felt electrodes. Journal of Power sources, 45(1), 29-41.
Zimniewska, M.; Michalak, M.; Krucinska, I.; Wiecek, B. 2003. Electrostatical and thermal properties of the surface of clothing made from flax and polyester fibres. Fibres and Textiles in Eastern Europe, 11(2), 55-57.
Zou, L.; Lan, C.; Li, X.; Zhang, S.; Qiu, Y.; Ma, Y. 2015. Superhydrophobization of Cotton Fabric with Multiwalled Carbon Nanotubes for Durable Electromagnetic Interference Shielding. Fibres and Polymers, 16(10), 2158-2164.